**A history of theory of structures
in the nineteenth century**

A history of theory of structures in the nineteenth century

T. M. CHARLTON

EMERITUS PROFESSOR OF ENGINEERING,
UNIVERSITY OF ABERDEEN

CAMBRIDGE UNIVERSITY PRESS

CAMBRIDGE

LONDON   NEW YORK   NEW ROCHELLE

MELBOURNE   SYDNEY

PUBLISHED BY THE PRESS SYNDICATE OF THE UNIVERSITY OF CAMBRIDGE
The Pitt Building, Trumpington Street, Cambridge, United Kingdom

CAMBRIDGE UNIVERSITY PRESS
The Edinburgh Building, Cambridge CB2 2RU, UK
40 West 20th Street, New York NY 10011–4211, USA
477 Williamstown Road, Port Melbourne, VIC 3207, Australia
Ruiz de Alarcón 13, 28014 Madrid, Spain
Dock House, The Waterfront, Cape Town 8001, South Africa

http://www.cambridge.org

First published 1982
First paperback edition 2002

A catalogue record for this book is available from the British Library

Library of Congress catalogue card number: 81-15515

ISBN 0 521 23419 0 hardback
ISBN 0 521 52482 2 paperback

Contents

Preface

The objective of this work is to provide an account, with appropriate detail, of some of the salient features of the development of the theory and analysis of engineering structures during the nineteenth century. There seemed to be two possible approaches to the subject: that whereby emphasis is on personalities and their contributions to the subject, or that whereby emphasis is primarily on subject development, with due acknowledgement of personalities and regard to the chronological aspect. Experience indicates that the former is conducive to some degree of repetition and confusion concerning the subject matter and, therefore, the latter approach is adopted (though personal names are used in subheadings to identify developments). But Chapter 6 is unique in being devoted to Levy's little-known, though highly-significant work on theory of frameworks. Free translation of original material is used extensively throughout the book in order to avoid misrepresentation or serious omission. Also, original notation for mathematical analyses are preserved as far as possible.

Chapters 4 and 11 differ from the others in being little more than brief reviews of topics which, though important, are peripheral to my purpose herein. The former embraces graphical analysis of simple frameworks which, together with the vast subject of graphical analysis of engineering problems generally, has very limited relevance to the features with which this work is concerned and which have determined the development of modern theory of structures. Then the Bibliography includes a number of references, in addition to those consulted, in order to provide as much guidance as possible to future research workers.

Appendices are devoted to Navier, Culmann and Robison respectively. The first is the generally acknowledged founder of modern theory of structures and elasticity; the second, one who exercised enormous influence on engineering analysis in the latter half of the century but whose (graphical) methods have fallen into disuse; and the third, an almost forgotten early contributor.

I am deeply indebted to Mrs Heather Flett and Mrs Margaret Rutherford for typing the manuscript, to Miss K. M. Svehla, B.Sc. Eng. for assistance with translation of material from the original German and to Mr Denis Bain for preparation of diagrams. I am also grateful to colleagues, especially Professor J. Heyman of the University of Cambridge and Sir Alfred Pugsley F.R.S., and to Cambridge University Press, for the encouragement without which the work would never have materialised.

T.M.C. 1981

1

Introduction

Much progress in theory of structures during the nineteenth century has been ascribed, notably by Clapeyron (1857) in France and Pole (Jeaffreson, 1864) in Britain, to the coming of the railway era. But the state of knowledge of the subject at the beginning of the century was ripe for rapid development due, for example, to Coulomb's remarkable research in applied mechanics. Early in the century Navier began to contribute to engineering science encouraged by his uncle, M. Gauthey, Inspector General of bridges and highways in France. Navier was born in 1785 and orphaned when he was fourteen years of age. He was adopted by Gauthey whose book on bridges he published (1809) and revised (1832), following his education at L'Ecole Polytechnique and then at L'Ecole des Ponts et Chaussées, from where he had become ingenieur ordinaire in 1808. He may be regarded as the founder of modern theory of elasticity and its application to structures and their elements. The year 1826 is memorable for the publication of Navier's celebrated *Leçons* as well as for the completion of Telford's remarkable wrought iron chain suspension bridge at Menai (it was also a year of sadness for Navier due to failure, prematurely, of the Pont des Invalides, a Paris suspension bridge which he had designed).

There is little doubt that France then led the world in the application of scientific principles to practical problems, having established, in 1784 and 1747 respectively, those two outstanding places of learning attended by Navier (l'Ecole Polytechnique was at first directed by Monge who was later joined by such outstanding figures as Fourier, Lagrange and Poisson: l'Ecole des Ponts et Chaussées was at first under the direction of Perronet who was a distinguished bridge engineer; later, Prony was its Director).

It is tempting to assert that the needs of engineering practice generated by, say, new materials, led to relevant advances in applied mechanics. Some major advances were, however, premature, apparently stimulated, at least in part, by the natural curiosity of individuals. Unfortunately, such premature advances seem to have received little attention and, being overlooked by practitioners, were forgotten, only to be rediscovered much later (and sometimes applied in an inferior manner), for example Navier's theory of statically-indeterminate bar systems.

Initially, masonry, timber and cast iron were the principal materials of construction, and their properties dictated the nature of structural forms: the arch to utilise the compressive strength of masonry or cast iron; the beam and latticework to utilise both the tensile and compressive strengths of timber. The manufacture and rolling of wrought iron (strong in tension and compression) was in its infancy in 1800 but it was to have a profound effect upon the theory and practice of construction.

In Britain the influence of the Rev. Professor Henry Moseley F.R.S. and his disciple Pole (later, professor at University College, London) on those renowned pioneers of railway construction, Robert Stephenson and Isambard Kingdom Brunel, is especially illuminating. (It shows, moreover, that contrary to widespread belief and Culmann's critical remarks, noted below, early outstanding advances in iron bridge construction in Britain were the result of using advanced scientific principles and experimental techniques as well as ingenuity of construction.) Moseley was noted for various contributions to engineering science and was familiar with French engineering science, especially the work of Coulomb, Navier and Poncelet. He was a pioneer of engineering education along with others including Robison, Willis and Rankine; but their influence does not seem to have been sufficient to achieve, in Britain, an enduring unity of theory and practice (the subject, incidentally, of Rankine's Inaugural Lecture to the University of Glasgow in 1856) such as that which was typified in the Britannia Bridge.

Moseley received his education in France as well as in Britain. He graduated in mathematics at St John's College, Cambridge and in 1831, when thirty years of age, he became professor of natural philosophy and astronomy at King's College, London, where he carried out research in applied mechanics, a subject which he taught to students of engineering and architecture 1840–2. In 1843 (the year before he left London to become, first a Government Inspector of Schools, and then, in 1853, a residentiary canon of Bristol Cathedral), his book *The mechanical principles of engineering and architecture*, which was based on his lectures, was published. It is probably the first comprehensive treatise on what might be called modern engineering mechanics to appear in English. There are acknowledgements to Coulomb (especially with regard to arches, earth pressure and friction), Poncelet (notably on elastic energy) and Navier (deflexion of beams and analysis of encastré and continuous beams). The appearance throughout of principles of optimisation (*extremum* principles relating to mechanical devices as well as statics) is noteworthy.

Moseley devoted much effort to a rigorous analysis of the stability of masonry voussoir arches (1835), from which emerged the concepts of line of pressure and line of resistance and the 'principle of least pressures'. (See Chapter 3: that principle has not survived but it inspired fruitful research by Cotterill.) But though his work on arches seems to explain observed modes of failure, it was to cause more confusion than enlightenment among British engineers, in spite of the interpretive efforts of Barlow (1846) and Snell (1846). Heyman has noted (1966) that the theory of the stability of a masonry arch due to Coulomb (1776) is unsurpassed. It is interesting, though, that Moseley concluded by vindicating the designs of arches by Rennie and others, based on the so-called 'wedge theory' of simple statics neglecting friction. The latter has been variously ascribed to Hooke, De La Hire, Parent and David Gregory (Charlton, 1976*b*). Indeed, that elementary theory was used throughout the century, notably by Brunel, without adverse consequences and independently of the ultimate strength theories of Coulomb and Villarceau.

It is for the design of continuous tubular wrought iron plate girder bridges (among the earliest continuous iron bridges in Europe) that Moseley's work is especially significant. In 1849 the theory of continuous beams, due to Moseley's teaching of Navier's methods, was used both for the Britannia Bridge and for the Torksey Bridge of John Fowler. This latter structure was continuous over three supports and, although it was very much smaller and altogether less enterprising than the Britannia, it was, nevertheless, of much interest (as noted in Chapter 2). It was completed nearly two years before the Britannia Bridge and declared unsafe for public use by the Government Inspector, Captain Simmons, R.E., in the atmosphere of suspicion of iron railway bridges which followed the Royal Commission of 1847–8 to inquire into the application of iron (railway) structures 'exposed to violent concussions and vibration' (Stokes, 1849). (That was after the fatal accident at Chester, for which Simmons was Inspector, when a trussed cast iron girder railway bridge by Stephenson collapsed while carrying a train.) The matter came to a head when Fairbairn, patentee of the tubular girder (1846), according to Pole (Jeaffreson, 1864), read a paper 'On tubular girder bridges' before the

Institution of Civil Engineers (1850). It was followed by a lively discussion which concentrated on the safety of the Torksey Bridge. Pole, Captain Simmons, Wild and Professor Willis were among the leading participants. Pole described his detailed analysis of the structure (Chapter 2), along with his results which supported the calculations and experiments of Wild for Fowler and vindicated the safety of the bridge. The Secretary of the Institution, Manby, strongly criticised government interference as being detrimental to progress. The contribution of Willis, Jacksonian Professor of Natural Philosophy at Cambridge, was concerned with the dynamic aspect especially. He was a member of the Royal Commission and distinguished for his experiments on the effect of a load travelling across a metal beam (1849). (His collaboration with Professor Stokes resulted in the latter's celebrated paper of 1849 and heralded the beginning of the precise study of dynamics of structures (Chapter 11), a subject which attracted Navier's attention with regard to suspension bridges.)

On the continent of Europe, theory of continuous beams (Chapter 2) was pursued, after Navier, notably by Clapeyron in 1848. Clapeyron, with Lamé, became professor at The Institute of the Engineers of Ways of Communication in St Petersburg in 1823 and, according to Timoshenko (1953), had a profound influence on the development of theory of structures in Russia. (Jourawski of that institute was engineer for the first railway bridges in Russia and developed theory of trusses as long ago as 1847, as well as exploiting continuous beams.) Among their early works was a memoir on the analysis of arches (1823), and Chalmers (1881) notes their work in graphical statics (1826b). Clapeyron is remembered popularly for his theorem of the three moments for continuous beams (1857), which resulted from the construction of the Pont d'Asnières near Paris. The idea was published first, though, by Bertot in 1855. Among the other contributors to aspects of continuous beam theory were (Chapter 2) Bresse, Mohr and Winkler; and it is important to acknowledge the method due to the elastician Clebsch, given in his celebrated book published in 1862 (and again, in 1883, in French, with annotations by Saint-Venant with Flamant). Subsequently, graphical methods (due to Mohr and especially Claxton Fidler) alleviated continuous beam analysis for some engineers.

It is interesting that Moseley and others seemed to overlook Navier's elegant method of analysing statically-indeterminate systems of bars (pin-jointed systems), a subject which was to be approached anew by Maxwell (1864b) and Mohr (1874a) as the need arose. Also, Clebsch included it in his book and developed it further. He treated space systems and introduced stiffness coefficients of linear elasticity with their reciprocal property (so, in a sense, establishing priority for the reciprocal theorem over Maxwell, Betti and Rayleigh, but he did not deal, it seems, with the physical implication). Another eminent elastician aware of the whole of Navier's *Leçons* was, of course, Saint-Venant who was responsible for the third (1864) edition. It seems that leadership in theory of elasticity and structures began to pass from France to Germany soon after that.

Navier had also provided the basis for elastic arch theory in his *Leçons* where the solution of the two-pin (statically-indeterminate) arch appears, for example. Bresse (1854) is, however, usually credited for this aspect having regard to his extensive treatment. It was at this time that Bresse noted and exploited the principle of superposition for linearly elastic structures (and linear systems generally) with regard to symmetry and anti-symmetry (Chapter 3). Culmann, Mohr, Winkler and Müller-Breslau were among later contributors to arch analysis and, according to W. Ritter (1907), it is to Culmann that the concept of the elastic centre is due, though Mohr seemed to identify the device (Chapter 10) with respect to a framed arch. Winkler's theory of stress in elements of large curvature (such as crane hooks) is noteworthy and he and Mohr *c.* 1870 suggested the use of theory of elasticity for analysing stone or masonry arches.

The economies apparently afforded by lattice girders and trusses in relation to arch and plate iron structures were being explored in the middle of the century, by which time mass production of wrought iron sections had begun. Those economies appeared to be related to the increase in the length of single, simply supported spans, which was afforded by the reduction in self-weight per unit length of such construction (in comparison with conventional beams) and to the inherent convenience of the beam as such. The flat-strip lattice girder was first known in Britain in the form of Smart's patent iron bridge, after Smart who, according to O. Gregory (1825), invented it in 1824. Its design was apparently based on the beam theory, the latticework being assumed equivalent to a plate web. These forms of construction were developed enthusiastically in the U.S.A., using timber as well as iron to build some impressive viaducts. Whipple achieved fame there for his truss designs based on sound principles, and his book published in 1847 contains, it is believed, one of the earliest thorough treatments of the analytical statics of complex frameworks (without redundant elements). The Russian Jourawski is said by Timoshenko (1950, 1953) to have initiated precise analysis of trusses, while Emmerson (1972) believes that Robison (professor of natural philosophy at Edinburgh) did so some fifty years earlier for timber trusses. Robison is also acknowledged by Cotterill (1884) and Weyrauch (1887) but Straub (1952) dismisses him as unsound. Whipple attempted to achieve true pin-jointing of his trusses, whereas the gusset plate and rivetting was used elsewhere, for example in Britain.

Both Moseley and Weale omitted the truss from their books (1843), indicating, apparently, that it was not then being used on a significant scale. Indeed, the theoretical content of Weale's *Theory, practice and architecture of bridges* (1843) deals mainly with masonry arches and suspension bridges and was contributed by Moseley and his colleague at King's College, London, W. Hann.

Although, in Britain, lattice girder railway bridges had appeared by 1844 (Hemans, 1844) the earliest truss for bridges seems to have been due to Captain Warren. According to Pole (Jeaffreson, 1864) it was used first for a major bridge (London Bridge Station) in 1850. Then in 1852 Cubitt used it for carrying a branch line of the Great Northern Railway over the Trent near Newark. The Warren–Kennard girder (de Maré, 1954) was used in the building of the spectacular Crumlin Viaduct (Ebbw Vale, Monmouth-shire) in 1857 and also in the Melton Viaduct (Okehampton). At about that time Bouch, assisted by Bow, was using the now so-called 'double' Warren girder for viaducts of the South Durham and Lancashire Union Railway, including the impressive Belah Viaduct. Indeed, Bow notes (1873) that these projects began in 1855. The viaducts consisted of a number of simply supported spans, and elementary statics was used for their design. Whewell's *Mechanics applied to the arts* (1834) and *The mechanics of engineering* (1841) exemplify the sound knowledge of elementary analytical statics available to engineers. (In the latter work, dedicated to Willis, the principle of using elastic properties to deal with statical indeterminacy is described briefly.)

The truss and its design provided impetus for the development of graphical methods of analysis. Bow (1873) recalls seeing a paper by Wild in 1854, which gave a complete graphical analysis of a simple truss (Bow, 1873, Fig. 243(*i*)), though he believed the date of the paper to be earlier. Graphical methods brought truss analysis within the competence of engineering draughtsmen and their origin is usually ascribed to Rankine (1858) and Maxwell (1864*a*), notwithstanding Bow's acknowledgement of Wild. But the use of graphical analysis in statics was not new. Varignon's funicular polygon (1725) and Coulomb's celebrated work in respect of earth pressures and masonry structures (Heyman, 1972) are noteworthy examples of the use of graphical analysis in the eighteenth century. Early in the nineteenth century, while he was a prisoner of war, Poncelet's interest in geometry for analytical purposes resulted in the creation of the new (projective) geometry (*c.* 1813), according to Chalmers (1881). Poncelet undoubtedly came to be regarded as a creator of modern engineering mechanics. Later, the initiative for extensive application of graphical methods was taken by Culmann at Zurich.

Culmann became professor of engineering sciences at the Zurich Polytechnikum in 1855 after experience in railway construction, and at a time of heavy demand on knowledge of theory of structures for the design of novel economical bridges on the truss principle. His highly distinguished contributions to graphical analysis, culminating in his celebrated book published in 1866, established his dominance of the new discipline called graphical statics which embraced engineering analysis generally. He based much of his work on the 'new geometry' of Poncelet (and Möbius). Curiously, he denied credit to Maxwell for the concept of reciprocal figures in framework analysis but Cremona (professor of mathematics at Milan and later at Rome), to whom he gave credit for it, acknowledged Maxwell's priority in a particularly lucid account of the subject (1872; see also Chapter 4).

Culmann's rejection of Maxwell probably reflected his disdain for British engineering (like Navier he visited Britain and the U.S.A., though nearly thirty years later, to study advances in bridge engineering which was at that time mainly related to railways). His opinions in this respect are recorded by Chalmers (Chapter 4), an ardent admirer of Culmann, who gives an extensive historical review in the preface of his scholarly book (1881). He quotes Culmann (from his book of 1866): 'But what is appropriate to the rich Englishman, who everywhere carries himself about with great consciousness, "I am in possession of the iron and do not require to trouble myself about statics", is not so to the poor devils of the Continent....' He contrasted the differences between the Continent and Britain with regard to the preparedness for Culmann's powerful methods, referring to the University and High School system in Germany where students were familiar with the works of Poncelet, Möbius and Chasles; while in Britain, the 'modern geometry' received little attention. Chalmers deplored the failure in Britain to accept the vital need for scientific training of engineers, thus:

There are, no doubt, among us, a large number who in earlier years have studied their Pratt, their Navier, their Moseley, or who in more recent years have become familiar with their Bresse and Rankine, have made themselves familiar with Clapeyron's Theorem of the Three Moments, even a few to whom Lamé is not unknown but those have done so without hope of reward.

Moreover, Weyrauch is quoted as asserting (1873) 'that continuous beams are popular only in countries where engineers can calculate' (a criticism of Britain which might well have been valid in 1880 but which neglected the priority attaching to the Britannia Bridge in both science and technology some thirty years earlier).

In 1881, Culmann died and was succeeded by Wilhelm Ritter, his former pupil and professor at the Riga Polytechnic Institute. Ritter published a major work on graphical statics (1888–1907). But his work should not be confused with that of August Ritter, Culmann's contemporary at Aix-la-Chapelle, whose book on theory and calculation of iron bridges and roofs (1862) was translated into English by Captain Sankey. In the course of that task, Sankey observed that Ritter's so-called method of moments (or sections) had already been discovered by Rankine (1858) when Ritter's book first appeared. (Schwedler is sometimes credited with the same idea.)

A distinguished contemporary of Culmann was Levy, in France. A pupil of Saint-Venant, he made a profound contribution to the modernisation of graphical statics in France, with his book, published in 1874, which contained *inter alia* some original matter of a purely analytical nature with regard to structures (Chapter 6). Williot's method of finding deflexions of trusses graphically (1877) is also a significant French contribution.

In Britain the degree of sophistication in graphical statics was less than on the Continent, as noted by Chalmers. Nevertheless, in addition to the contributions of Maxwell (1864a), Jenkin (1869), and Bow (1873) who was famed for his notation for force diagrams of trusses, there were the distinguished contributions of Claxton Fidler's analysis by 'characteristic points' of continuous beams (1883) and Fuller's graphical method for arches (1874). (Moreover, Maxwell was, after Jenkin's encouragement in 1861, also concerned simultaneously with the analysis of statically-indeterminate frameworks.)

On the Continent, Castigliano in Italy (1873), Levy in France (1874) and Mohr in Germany (1874a) had achieved priority in various ways of analysing statically-indeterminate trusses or bar frameworks, in addition to devoting much attention to graphical analysis where appropriate. Levy, mindful of Navier's method published in his *Leçons* some fifty years earlier, suggested an alternative approach, while Mohr appears to have attacked the problem *de novo* in a manner essentially the same as that of Maxwell (the details of these methods are described in Chapters 5 and 6). Castigliano, with full knowledge of Navier's method (like Levy) sought (as described in Chapter 8) an alternative based on elastic energy derivations (after Menebrea's unsuccessful attempt of 1858). Castigliano gives, incidentally, an account of the method due to Navier in his book published in 1879 (it seems that Navier's *Leçons* were translated into Italian). Incidentally, Castigliano, like so many other leaders of structural analysis in Europe (Clapeyron, Culmann, Jourawski, Engesser, Mohr, Rankine, Jasinsky, Winkler and Crotti), was concerned with railway construction.

Having (like Mohr and Levy, after Culmann) developed sophisticated graphical methods for dealing with a wide variety of problems (of which truss analysis was only one) Müller-Breslau turned his attention eagerly to promoting and extending the new analytical methods of Maxwell–Mohr and Castigliano. (Mohr's very different attitude to both Castigliano and Maxwell is described in Chapter 10.) Müller-Breslau's major work on graphical statics (1887*b*) includes much that is not concerned with graphical or geometrical methods (after the manner of Levy) and adds a high degree of sophistication and clarity to the analysis of statically-indeterminate frameworks. Thus, he appears to have introduced uniformity by the notation of flexibility (influence coefficients) for linearly elastic structures and so to have relieved the formulation of the solution, for any type of structure, from any particular method of calculating deflexions (that is, the flexibility coefficients would be calculated in a manner which depended upon individual preference). Müller-Breslau is popularly remembered for the theorem regarding influence lines for forces in elements of statically-indeterminate structures, which bears his name (Chapter 10). He is also credited (as is Southwell) with the concept of tension coefficients for space frame analysis, but the concept is implied by Weyrauch (1884).

In the meantime Engesser, and Castigliano's friend Crotti were adding to knowledge of energy principles in theory of structures. (Castigliano's theorems of strain energy, especially his so-called 'principle of least work', for relieving analysis of statically-indeterminate structures of conceptual, physical thought, quickly received widespread acclaim on the Continent and in Britain later, due to Martin (1895) and Andrews (1919).) The origins of energy methods in theory of structures or practical mechanics may be traced to Poncelet (Chapter 7) and Moseley. (Part 2 of Castigliano's first theorem of strain energy was anticipated by Moseley (1843).) Indeed, Moseley's disciple, the mathematician Cotterill, anticipated the essence of Castigliano's principle of least work in 1865 in the *Philosophical Magazine*, which escaped the notice of engineering scientists in Europe and, later, Fränkel (1882) discovered it independently. Apart from their philosophical interest, involving speculation regarding economy of Nature, the *extremum* principles resulting from research into energy concepts were to be valuable, mainly for obtaining rapid approximate solutions to certain complicated problems.

Castigliano's least work theorem was among the first methods used for rigidly-jointed frameworks, both with regard to portals and trusses, and elastic theory of suspension bridges. It featured, for example, notably in Müller-Breslau's book *Die neueren Methoden* (1886*b*) which represented a major advance in the literature of theory of structures.

The fear of failure of major bridge trusses, due to the stresses induced by the rigidity of joints (secondary stresses), stimulated research by a number of distinguished German engineers after 1877. In that year, according to Grimm (1908), a prize was offered by the Polytechnikum of Munich for the solution of that problem. The term *Sekundärspannung* (secondary stress) was, it seems, originated by Professor Asimont of that institution, to distinguish between the direct and bending-stresses in an eccentrically loaded column. Asimont formulated the problem with regard to rigidly-jointed trusses and Manderla's solution (1879) gained him the prize, it appears. Before the publication of Manderla's solution in 1880, however, Engesser had published an approximate method. Also, Winkler indicated, in a lecture on the subject (1881), that he had given attention to it for some years past. In 1885 Professor Landsberg contributed a graphical solution which was followed by another analytical solution by Müller-Breslau in 1886 (1886*a*). Another graphical solution appeared, it seems, in 1890, this time from W. Ritter; and then in 1892 a further analytical solution was contributed by Mohr. Engesser published a book on the analytical determination of secondary and additional stresses in 1893. These aspects are considered in more detail in Chapter 11.

Although Müller-Breslau used Castigliano's energy method to analyse a simple (single storey, single bay) rigidly-jointed portal framework, that kind of structure did not, it seems, attract the degree of attention given to secondary effects in trusses until the twentieth century. The appearance of Vierendeel's novel design for open-panel bridge girders in 1897 (according to Salmon, 1938) posed a formidable analytical problem for which approximate methods (for example the use of estimated points of contraflexure) were used initially.

This review of nineteenth-century structural engineering, with reference to theory of structures, would be incomplete without mention of the problems posed by major suspension bridges. In spite of adverse experience, particularly in Britain, the Americans persevered and in 1855 adopted the suspension principle successfully for a major railway bridge to Roebling's design, over the Niagara Falls. Although, at that time, theory of suspension bridges (Chapter 3) was based on simplifying assumptions to render the problem amenable to statics alone (as, for example, in

Rankine's theory, 1858) and, moreover, the provision of ties from deck to towers to supplement the cables was common, it seems the concept of gravity stiffness was being recognised implicitly, if not explicitly. Thus, Pugsley (1957, 1968) quotes from a letter Roebling sent to the company prior to the building of the Niagara Bridge: 'Weight is a most essential condition, where stiffness is a great object.' Then in 1883 the great Brooklyn Bridge was built to Roebling's design, judged by Pugsley to be a triumph of intuitive engineering. The completion of this bridge almost coincided with the origins of elastic theory to which W. Ritter (1877), Fränkel (1882), Du Bois (1882) and Levy (1886) were early contributors (Chapters 3 and 9), and in which elastic deflexion of the deck, due to live load, determined the uniformly distributed reaction provided by the cables, as dictated by compatibility of displacements. Approximation regarding small deflexion of the cables was involved and substantial improvement in this respect, following the introduction of gravity stiffness by Melan (1888), whereby accurate analysis of bridges with relatively flexible decks (which depended greatly on that source of stiffness), became possible. The modern deflexion theory of suspension bridges is due essentially to Melan and then to Godard (1894).

In conclusion it is interesting to recall Pole's commentary (Jeaffreson, 1864) on iron bridges, with emphasis on British practice. He remarks that the history of iron bridges commenced in the sixteenth century when such structures were proposed in Italy and then that an iron bridge was partly manufactured at Lyons in 1755, but that it was abandoned in favour of timber in the interests of economy. In the event, however, the first iron bridge (of cast iron) was erected in Britain, being completed in 1779. The builder was Abraham Darby and the site was the River Severn at Coalbrookdale, Shropshire. But Pole notes that as early as 1741 a wrought iron-chain footbridge was erected over the River Tees, near Middleton, County Durham. That kind of bridge was later developed by Captain Samuel Brown (for example the Union Bridge of 1819 near Berwick) who used long iron bars instead of ordinary link chains. (In the U.S.A., Finlay built an iron suspension bridge in 1796 at Jacob's Creek, and Séguin's bridge at Ainé was built in 1821.)

Pole notes the impetus to bridge engineering, given by the coming of railways, and the need to discover an alternative to the arch for a variety of circumstances. He suggests that the simple beam ('the earliest form of all') made of iron, afforded the desired economical solution with its possibilities for development. The 'five great properties' claimed for it were:

    1 rigidity;
    2 convenience with regard to being level or straight (for example the Britannia Bridge);
    3 simplicity of abutment conditions (that is, no horizontal thrust);
    4 the ironwork for a beam is less in weight than for an arch of the same strength and span;
    5 convenience of erection and less interference with navigation during construction over water courses.

Pole summarises iron bridges as belonging to three classes: the iron arch; the suspension bridge; and the iron girder bridge, with the last of greatest utility in the eight categories:
    1 solid beams;
    2 trussed cast iron girders;
    3 bowstring girders (cast iron, as in the High Level Bridge by R. Stephenson at Newcastle; wrought iron at Windsor, by Brunel);
    4 simple 'I' girders;
    5 tubular or hollow plate girders (e.g. Britannia Bridge);
    6 triangular framed girders (including the Warren girder);
    7 lattice girders;
    8 rigid suspension girders (Brunel's Chepstow Bridge and Saltash Bridge).

The remainder of Pole's article is devoted to specific instances of the use of the various kinds of iron bridge and the failure of some of them, notably Stephenson's trussed cast iron girder bridge at Chester in 1847.

Notes

Robison contributed articles on applied mechanics to the *Encyclopaedia Britannica* (1797) and Brewster (1822) refers to them and to Young's version of them in the next edition.

Straub (1952) and Timoshenko (1953) provide details of the foundation of l'Ecole Polytechnique and l'Ecole des Ponts et Chaussées.

Straub (1952) observes that the absence of graphical methods in Navier's *Leçons* is in contrast with 'modern' text-books, an observation of questionable validity even thirty years ago.

According to Dempsey (1864), Wild joined Clark as assistant to Robert Stephenson on the Britannia Bridge.

Rankine's Inaugural Lecture to the University of Glasgow is reproduced in his *Manual of applied mechanics* (1858).

Clapeyron and Lamé designed iron suspension bridges (1826*a*) which were constructed during the years 1824–6 in St Petersburg and were among the first on the continent of Europe.

The concept of influence lines seems to have arisen (1868) through Winkler's work in relation to elastic arches (Chapter 3).

Mosely and Fairbairn were distinguished as the only British Corresponding Members in Mechanics of the French Academy in 1858, the year of Clapeyron's election in preference to Saint-Venant and three others, to fill the vacancy due to Cauchy's death. In the same year the Academy appointed Clapeyron to a Commission to advise on the Suez Canal project.

Cotterill (1869) brought to the attention of British engineers the use of the funicular polygon for the graphical calculation of bending moments of simple beams. He referred to a preliminary version of Culmann's *Die graphische Statik* (1866) published in Leipzig in 1864 and to Reuleaux (1865).

2

Beam systems

With the theories of flexure and bending-stress in beams, established in the eighteenth century by James (Jacob) Bernoulli and Euler (*c.* 1740) and Coulomb (1773) respectively, Navier developed the analysis of forces and deflexions of beams of varying degrees of complexity, with regard to support and restraint, as part of his extensive and unique researches in theory of elasticity. In those researches, evaluated by Saint-Venant and others (1864), he laid the foundations of modern technical theory of elasticity and anticipated important applications.

It had become well known in carpentry that continuity of beams over supports and building-in the ends of beams, contributed substantially to their strength or carrying capacity. Indeed, Robison had considered this subject in an elementary fashion toward the end of the eighteenth century (Brewster, 1822). Navier was clearly mindful of the common use of such statically-indeterminate construction in timber (to judge by the detail of his illustrations) when he embarked on the precise analysis of systems of that kind and, in the event, his analysis was timely with regard to the development of wrought iron beams and structures, which was stimulated by the needs of railway construction. It was, in fact, the statically-indeterminate beam (including, especially, the continuous beam) which dominated the development of the beam in the nineteenth century.

Navier, 1826

The analysis of encastré and continuous beams is believed to have been published for the first time in Navier's celebrated *Leçons* of 1826 (though Clapeyron refers to earlier lithographed notes). Saint-Venant claimed (1883) to have given his own version of the analysis of beam systems in lithographed lectures delivered in 1838, without attempting to detract from Navier's priority. Indeed, he edited the third and final edition of Navier's *Leçons* (1864) to include a résumé of the history of elasticity and strength of materials as well as biographical details and tributes to Navier by Prony and others.

Having used the differential equation of the elastic line to find the deflexion of a uniform, simply supported beam for both concentrated and distributed loads, Navier continues by analysis of the (statically-indeterminate) propped cantilever and the encastré beam. He remarks (like Robison) that a uniform encastré beam is twice as strong as the same beam simply supported for a load at mid-span. Then he turns his attention to a uniform beam with three or more points of support using the diagram shown in Fig. 1 (his fig. 50, 1826) for illustration. The precise nature of his analysis of beams is conveniently illustrated by this example.

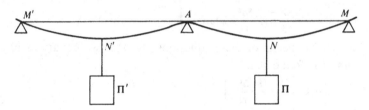

Fig. 1. From Navier (1826).

The beam $M'M$ of length $2a$ is supported at M' and M and its mid-point A, as shown. There are concentrated loads Π at N, distant $a/2$ to the right of A; and Π' at N', $a/2$ to the left of A. The reactions to these loads are denoted by p, q, q' at A, M and M' respectively; and ω is the consequent angle of slope at A. For equilibrium with regard to forces:

$$\Pi + \Pi' = p + q + q' \qquad\qquad [(2.1)]$$

and, by taking moments about A:

$$\Pi - \Pi' = 2(q - q') \qquad\qquad [(2.2)]$$

Assuming the span AM is encastré at A, for the portion AN:

$$\epsilon \frac{\mathrm{d}^2 y}{\mathrm{d}x^2} = \Pi\left(\frac{a}{2} - x\right) - q(a - x) \qquad\qquad [(2.3)]$$

$$\epsilon \frac{\mathrm{d}y}{\mathrm{d}x} = \Pi\left(\frac{ax}{2} - \frac{x^2}{2}\right) - q\left(ax - \frac{x^2}{2}\right) + \epsilon \tan \omega \qquad\qquad [(2.4)]$$

$$\epsilon y = \Pi\left(\frac{ax^2}{4} - \frac{x^3}{6}\right) - q\left(\frac{ax^2}{2} - \frac{x^3}{6}\right) + x\epsilon \tan \omega \qquad\qquad [(2.5)]$$

Also, for the portion NM, using the condition that at N, where $x = a/2$, dy/dx and y are the same as for AN

$$\epsilon\frac{d^2y}{dx^2} = -q(a-x) \qquad\qquad [(2.6)]$$

$$\epsilon\frac{dy}{dx} = -q\left(ax - \frac{x^2}{2}\right) + \Pi\frac{a^2}{8} + \epsilon\tan\omega \qquad\qquad [(2.7)]$$

$$\epsilon y = -q\left(\frac{ax^2}{2} - \frac{x^3}{6}\right) + \left(\Pi\frac{a^2}{8} + \epsilon\tan\omega\right)x - \Pi\frac{a^3}{48} \qquad\qquad [(2.8)]$$

Navier then says that the corresponding equations for portions AN' and $N'M'$ are similar except that Π' is written instead of Π, q' instead of q and the sign of $\tan\omega$ is changed. Also, since for both portions $y = 0$ when $x = a$:

$$\left.\begin{array}{l} 0 = -q\dfrac{a^2}{3} + \Pi\dfrac{5a^2}{48} + \epsilon\tan\omega, \\[3mm] 0 = -q'\dfrac{a^2}{3} + \Pi'\dfrac{5a^2}{48} - \epsilon\tan\omega. \end{array}\right\} \qquad\qquad [(2.9)]$$

Finally, by means of these equations and the two equations of equilibrium, it is found that:

$$\left.\begin{array}{l} \tan\omega = \dfrac{\Pi - \Pi'}{\epsilon}\cdot\dfrac{a^2}{32} \\[3mm] p = \dfrac{22\Pi - 22\Pi'}{32} \end{array}\right\} \qquad\qquad [(2.10)]$$

$$\left.\begin{array}{l} q = \dfrac{13\Pi - 3\Pi'}{32} \\[3mm] q' = \dfrac{3\Pi + 13\Pi'}{32} \end{array}\right\} \qquad\qquad [(2.11)]$$

He specifies that rupture will take place first at A, N or N' and, having expressed the bending moments at those points ($\epsilon d^2y/dx^2$), proceeds to obtain expressions for the stresses.

The combination of analysis and physical insight of Navier's work is remarkable. (In his analysis of beams which are statically-indeterminate he uses conditions of compatibility of deflexion (and slope) for his final equations, having first made use of the relevant conditions of equilibrium.) It is worth noting that his method of dealing with a simply supported beam with a concentrated load is to use the point of application of the load as origin for dealing with the portions of the beam on either side, as cantilevers. The constants of integration are arranged so that the condition for continuity at the load are assumed.

Moseley, 1843

In his presentation (1843) of Navier's methods (the first in Britain), Moseley begins by dealing with the theory of bending used by Navier. It is interesting that he demonstrates the calculation of the second moment of area of a beam section for a cast iron beam with unequal flanges: otherwise his illustrations of loaded beams indicate timber loaded by masonry. Apparently to simplify the mathematical detail, Moseley uses uniform beams with uniformly distributed loads exclusively (culminating in a beam supported at its ends and two intermediate points, in article 376), until he deals with 'conditions of the equilibrium of a beam supported at any number of points and deflected by given pressures' (the term 'pressure' refers to a concentrated force or load) on p. 521 of his *Mechanical principles of engineering and architecture*. He considers a uniform beam *CEBDA* (Fig. 2) simply supported at C, B and A with reactions P_5, P_3 and P_1, caused

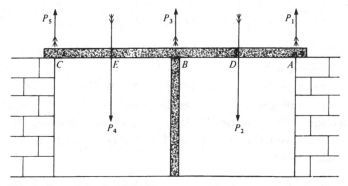

Fig. 2. From Moseley (1843).

by downward loads P_4 at E and P_2 at D. $CB = a_2$, $CE = a_2/2$; $BA = a_1$, $BD = a_1/2$. Moseley gives the differential equation of the neutral line between B and D as:

$$EI\frac{d^2y}{dx^2} = P_2(\tfrac{1}{2}a_1 - x) - P_1(a_1 - x) \qquad (568) \quad [(2.12)]$$

with origin at B and using the same sign convention as Navier for bending moment; it is noted, however, that now flexural rigidity is denoted by EI instead of ϵ.

Then for the portion DA:

$$EI\frac{d^2y}{dx^2} = -P_1(a_1 - x) \qquad (569) \quad [(2.13)]$$

Now 'representing by β the inclination of the tangent at B to the axis of the abscissae' and integrating the former equation twice between the limits 0 and x:

$$EI\frac{dy}{dx} = \tfrac{1}{2}P_2(a_1 x - x^2) - P_1(a_1 x - \tfrac{1}{2}x^2) + EI \tan \beta \qquad (570) \quad [(2.14)]$$

$$EIy = \tfrac{1}{2}P_2(\tfrac{1}{2}a_1 x^2 - \tfrac{1}{3}x^3) - \tfrac{1}{2}P_1(a_1 x^2 - \tfrac{1}{3}x^3) + EIx \tan \beta \qquad (571) \quad [(2.15)]$$

At $x = a_1/2$ (that is, at D) where $y = D_1$, and representing by γ the 'inclination to the horizon' of the tangent at D:

$$EI \tan \gamma = \tfrac{1}{8}P_2 a_1^2 - \tfrac{3}{8}P_1 a_1^2 + EI \tan \beta \qquad (572) \quad [(2.16)]$$

$$EID_1 = \tfrac{1}{24}P_2 a_1^3 - \tfrac{5}{48}P_1 a_1^3 + \tfrac{1}{2}EIa_1 \tan \beta \qquad (573) \quad [(2.17)]$$

Integrating equation (569) between the limits $a_1/2$ and x:

$$EI\frac{dy}{dx} = -P_1(a_1 x - \tfrac{1}{2}x^2) + EI \tan \gamma + \tfrac{3}{8}P_1 a^2 \qquad [(2.18)]$$

Eliminating $\tan \gamma$ between this equation and equation (572) and simplifying gives:

$$EI\frac{dy}{dx} = -P_1(a_1 x - \tfrac{1}{2}x^2) + EI \tan \beta + \tfrac{1}{8}P_2 a_1^2 \qquad (574) \quad [(2.19)]$$

Integrating again between the limits $a_1/2$ and x and using equation (573) to eliminate D_1 gives:

$$EIy = -\tfrac{1}{2}P_1(a_1 x^2 - \tfrac{1}{3}x^3) + (EI \tan \beta + \tfrac{1}{8}P_2 a_1^2)x - \tfrac{1}{48}P_2 a_1^3 \qquad (575) \quad [(2.20)]$$

Moseley continues: 'Now it is evident that the equation to the neutral line in respect of the portion CE of the beam will be determined by writing in the above equation P_5 and P_4 for P_1 and P_2 respectively' (and, presumably, a_2 and a_1).

Making this substitution in equation (575) and writing $-\tan \beta$ for $+\tan \beta$ in the resulting equation; then assuming $x = a_1$ in equation (575), and $x = a_2$ in the equation thus derived from it, and observing that y then becomes zero in both, we obtain:

$$\left. \begin{array}{l} 0 = -\tfrac{1}{3}P_1 a_1^3 + \tfrac{5}{48}P_2 a_1^3 + EIa_1 \tan \beta \\[4pt] 0 = -\tfrac{1}{3}P_5 a_2^3 + \tfrac{5}{48}P_4 a_2^3 - EIa_2 \tan \beta \end{array} \right\} \qquad [(2.21)]$$

Also, by the general conditions of the equilibrium of parallel pressures:

$$\left. \begin{array}{l} P_1 a_1 + \tfrac{1}{2}P_4 a_2 = P_5 a_2 + \tfrac{1}{2}P_2 a_1 \\[4pt] P_1 + P_3 + P_5 = P_2 + P_4 \end{array} \right\} \qquad [(2.22)]$$

Eliminating between these equations and the preceding, assuming $a_1 + a_2 = a$, and reducing, we obtain:

$$P_1 = \frac{P_2 a_1(8a_2 + 5a_1) - 3P_4 a_2^2}{16aa_1} \qquad (576) \quad [(2.23)]$$

$$P_5 = \frac{P_4 a_2(8a_1 + 5a_2) - 3P_2 a_1^2}{16aa_2} \qquad (577) \quad [(2.24)]$$

$$P_3 = \tfrac{1}{2}\left\{ P_2\left(1+\frac{3a_1}{8a_2}\right) + P_4\left(1+\frac{3a_2}{8a_1}\right) \right\} \qquad\qquad (578) \quad [(2.25)]$$

He continues and obtains expressions for deflexions D_1 and D_2 and the slopes, and then considers the special circumstances when $P_2 = P_4$ and $a_1 = a_2$. In a footnote he describes an experiment on a uniform wrought iron bar on three simple supports, performed by Hatcher at King's College, London (Fig. 3).

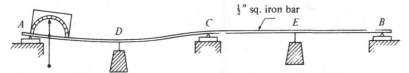

Fig. 3. Hatcher's continuous beam experiment.

Another topic which Moseley treats (unlike Navier) is the 'work expended upon the deflexion' of a uniform beam (articles 368, 369 and 385), which is denoted by the symbol U and, moreover, he makes use of the derivative $P = dU/d\Delta$ (Chapter 7). (It is believed that the concept of strain energy was not introduced into engineering mechanics until 1831 by Poncelet, whose profound influence on Moseley is evident in numerous references.)

Continuous tubular bridges, 1850

The theory of continuous beams was applied significantly and, in retrospect, uniquely, by British engineers *c*. 1850 (with the help of Moseley and Pole and others), especially for the design and erection of the Britannia Bridge of Robert Stephenson, over the Menai Straits, shown in Fig. 4 (which, according to Todhunter & Pearson (1886, vol. 1, para. 1489) along with the Conway Bridge comprised 'probably the most important problems to which the Bernoulli–Eulerian theory was ever or ever will be, applied'). Fortunately, details of its design and construction (with that of the single-span Conway Bridge) are well documented. Thus, Fairbairn published 'An account of the construction of the Britannia and Conway tubular bridges' in 1849 and in the following year Clark published a two-volume work *Britannia and Conway tubular bridges* under the supervision of his employer and engineer for the bridges, Robert Stephenson. (Fairbairn was employed by Stephenson for his expertise in wrought iron construction and design of tubular girders. Before the Britannia Bridge was finished they quarrelled and Fairbairn hastened to publish a book about the projects, which, in fact, appeared in the year before the Britannia Bridge

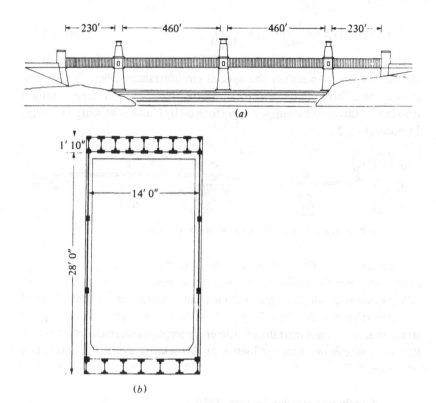

Fig. 4. (*a*) Britannia tubular bridge – elevation. (*b*) Britannia tubular bridge – section of tube.

was completed.) In so far as theory of structures is concerned, Clark's book contains an extensive account of the theory of bending of beams using 'the principles and methods of calculation laid down in Moseley's *Mechanical principles of engineering and architecture*'. Thus, in volume 1, chapter 2 is devoted to the derivation of formulae for the strength of beams; chapter 3 to deflexion of beams; and chapter 4 to the theory of continuous beams (including reference to the Torksey tubular bridge calculations by Pole). In chapter 6 there are details of experiments using wooden rods (including one 'communicated by Mr Brunel') to verify the results given by calculation in chapter 4 for continuous beams on four and five supports (Fig. 5). Then in volume 2, chapter 3, the strength and deflexion of the Britannia Bridge is considered in detail, including the method of equalising bending moments due to self-weight at mid-span and support points. It is this chapter which is especially significant in the history of theory of structures.

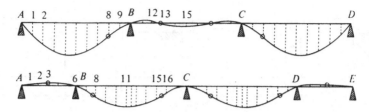

Fig. 5. Continuous beam experiment due to Brunel.

Pole, who was Fairbairn's biographer, as well as the leading protagonist of Moseley's methods and adviser for their application, is believed to be mainly responsible for it (as for the chapters on beam theory in the first volume). Two interesting details which emerge are the method used to observe the surface strains of the tubes at towers; and the assertion that a gale would cause a force of 20 lb wt per square foot on the side of a tube. The former consisted simply of measuring change in length by iron rods (20–50 ft long) attached to the surface.

The method of equalising extreme self-weight bending moment values, as essential for the safety of the structure in accordance with the design concept, apparently involved ingenious measurements during erection as well as calculations using theory of continuous beams (although the text of Clark's book is somewhat vague in respect of whether the calculations were carried out subsequently to verify procedures adopted by judgement at the site, or beforehand: the former seems the more probable). The bridge had two side spans of 230 ft and two main spans of 460 ft. Indeed, it was a pair of tubular bridges side by side, each of which was a rectangular tube through which one of the two parallel railway lines passed. It is sufficient to consider one complete tube with its four spans. The tube was manufactured in four sections corresponding to the spans, in a construction works close to the site of the bridge. Each section was raised from pontoons in the channel by jacking until it was in position, when permanent supporting stonework was completed. At a stage at which all four sections were in place on the supports, but before the joints between them over the intermediate supports had been completed, it was as though a continuous tube in an unstressed condition had been laid on the five supports and then severed over the three intermediate supports. This stage is shown in Fig. 6 which is taken from Clark's book.

The angles of separation between the four sections of the tube were observed with the aid of plumb lines. In order to achieve the degree of effect of continuity required, the procedure used was first to raise the end *E* of

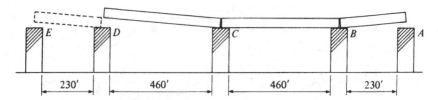

Fig. 6. From Clark (1850).

the side span *DE* until the angle of separation at *D* was almost closed, then to complete the joint between *CD* and *DE*. Next, the section *BC* was raised (*E* having been lowered to its normal level) to partially close the angle of separation which was observed at *C* (being less than the original amount, due to the completion of joint *D* and lowering of *E*). Finally, *A* was raised by precisely the same amount as *E* had been raised initially and the joint at *B* was completed. Thus the effect was to produce a continuous tube whose maximum bending moment, due to self-weight, was less than that for the unstressed continuous tube placed on the five supports. The ideal sought was equalisation of the bending moment at the central support and the extreme value within the adjacent main spans (due to self-weight). It appears that *E* was raised by an amount corresponding to the slope at *D* of the section *CD* as simply supported, so neglecting the bending, due to self-weight, of section *DE*. Raising *A* by the same amount before completing the final joint at *B*, as described by Clark, would apparently result in over-closure of the angle of separation at *B* (because it would in any event be reduced by the making of the joints at *D* and *C*). The effect of that would be to increase the bending moments at the supports, whereas the intention was stated as being to reduce them in relation to those which would correspond with exact closure.

The theoretical investigation confirms the amount by which *E* was raised, neglecting bending of *DE*, before the joint at *D* was completed. Then the rotation at *D* and the modified angle of separation at *C*, caused by lowering *E* after making joint *D*, are calculated by considering *CDE* as a continuous beam as follows:

$$\text{span } CD: \quad EI\frac{\mathrm{d}^2 y}{\mathrm{d}x^2} = \frac{\mu x^2}{2} - p_3 x \qquad [(2.26)]$$

where *I* and μ are the second moment of area and weight per unit length of *CD*, and p_3 is the reaction at *C*.

Integrating, and remembering that at the point D, where $x = 2l$, $\dfrac{dy}{dx} = \tan \beta$, we have –

(LXVII) $\quad EI\dfrac{dy}{dx} = \dfrac{\mu}{6}x^3 - \dfrac{p_3}{2}x^2 - \dfrac{4}{3}\mu l^3 + 2p_3 l^2 + EI \tan \beta$ $\hspace{2cm}$ [(2.27)]

Integrating again, and substituting in the resulting equation the corresponding values of $x = 2l$ and $y = 0$, we obtain –

(LXVIII) $\quad EI \tan \beta = \mu l^3 - \tfrac{4}{3}p_3 l^2$ $\hspace{5cm}$ [(2.28)]

By a similar process we obtain for the small span –

(LXIX) $\quad EI_2 \tan \beta = \tfrac{1}{3}p_1 l^2 - \tfrac{1}{8}\mu_2 l^3$ $\hspace{4.5cm}$ [(2.29)]

Now by equality of moments round D, we have –

(LXX) $\quad p_1 l + 2\mu l^2 = 2p_3 l + \tfrac{1}{2}\mu_2 l^2$ $\hspace{5cm}$ [(2.30)]

Eliminating, therefore, between the three last equations, we find the following values –

$$p_1 = 53 \text{ tons}$$
$$p_3 = 653 \text{ tons}$$
$$\tan \beta = 0.002 \text{ lb}$$

The values obtained relate to $2l = 2DE = CD = 460$ ft; $\mu = 3.38$ tons per ft; μ_2 (for DE) = 2.6 tons per ft; $I = 1584$ (ft)^4; and I_2 (for DE) = 962 (ft)^4.

It is interesting to note that no account is taken of the fact that in making the joint at D the original angle of separation was not completely eliminated, because the bending of DE due to its self-weight was neglected.

The value of the angle of separation at C obtained from this analysis is used to determine the height through which BC should be raised at B for complete closure and it is noted that it was actually raised by little more than half that amount when the joint at C was completed. That was believed to be satisfactory because 'the strain over the support at C is to that in the centre of the long span as 6 to 3' if full advantage of continuity is taken.

Finally,

The third and last junction made was that at B: the calculation of this would be very complicated, but we are justified in assuming that the tube was made perfectly continuous at this point...therefore it appears that the tube may be considered as made continuous at B and D, but falling short of perfect continuity at C by a certain known amount.

That statement is not strictly correct both with regard to the restoration of the conditions which full continuity would have afforded at B and D and with reference to the conditions at C being modified by 'a certain known amount'. The latter seems to have been a judicious estimate. Having regard, however, to the orders of magnitude involved and strains measured, it seems probable that the engineers were justified in their conclusions. (The achievement of the effect of perfect continuity could clearly have been attained entirely by site measurement, whereby the angles of separation would have been eliminated entirely at *D*, *C* and *B* in succession, by appropriate jacking and rigid-jointing.) Certainly, they seem to have adopted a sound procedure on a scientific basis to which the long life of the bridge in increasingly severe conditions is adequate testimony, in spite of subsequent criticism by Clapeyron (1857) and others.

Clapeyron's method, 1848–1857

It was Clapeyron, however, who generalised and ordered the analysis of continuous beams to advantage for practising engineers. Thus Bresse (1859) writes:

Beams on more than two supports have been used often in recent years for bridges constructed for railways. One of the first examples of this art in France is the bridge of the West railway over the Seine at Asnières. For the Asnières Bridge M. Clapeyron, Ingénieur en Chef des Mines, member of the Academy of Sciences, has greatly improved on the methods of verifying the strength of beams or for determining their proportions.

In a brief discourse to the Academy of Sciences and reported in *Comptes Rendus* (1857), Clapeyron reviewed the development of continuous girder bridges, ascribing it to the immense capital outlay in railway construction and giving Stephenson's Menai (Britannia) Bridge 'of originality and grandeur' as a prime example. After outlining briefly the origin of the theory of continuous beams and applications, with acknowledgement to a paper by Navier in *Bulletin de la Société Philomathique* (1825), and the subsequent studies of Belanger and Molinos & Pronnier, he described how he himself became involved in the problem for the first time in 1848 (for the Ponts d'Asnières) and his desire to submit his results to the judgement of the Academy. He then simply quoted the principal result as the equation:

$$l_0 Q + 2(l_0 + l_1)Q_1 + l_1 Q_2 = \tfrac{1}{4}(p_0 l_0^3 + p_1 l_1^3) \qquad [(2.31)]$$

where l_0 and l_1 are the lengths of two consecutive spans; Q_0, Q_1, Q_2 are the respective bending moments over the three consecutive supports; and p_0 and p respectively, are the uniformly distributed loads on the spans. If the spans are of equal length the equation is:

$$Q_0 + 4Q_1 + Q_2 = \frac{l^2}{4}(p_0 + p_1) \qquad [(2.32)]$$

Thus the 'theorem of the three moments' emerged.

Clapeyron proceeded to describe briefly the results of a study of the

Britannia ('Meny') Bridge by Molinos & Pronnier 'in which they have found that the iron bears a stress in the centre of the first span, of approximately 300 kg per square cm; over an outer pier, of 900 kg per square cm; at the centre of the second span, 550 kg per square cm; and 860 kg per square cm over the central pier'. Clapeyron remarked: 'This magnificent structure leaves much to be desired with regard to the distribution of thicknesses of the plate which seems relatively too thin at the points of support.'

The remainder of the discourse was devoted to a demonstration of the use of his equation for a beam of seven equal spans and means of solving the resulting simultaneous equations.

Clapeyron had apparently made his results known informally after 1848 and, as a consequence, in 1855 Bertot achieved priority in publishing the theorem of the three moments and, though he acknowledged Clapeyron's investigation, he was accorded priority by some, including Collignon and Heppel (as noted below). It is perhaps significant that Clapeyron ignored Bertot in his discourse before the Academy. In the meantime Jourawski was developing theory of continuous beams for railway construction in Russia, with some knowledge of Clapeyron's work, it is believed.

The Belgian, Lamarle (1855), seems to have made a trivial contribution to theory of continuous beams; then Rebhann (1856) used Navier's method and, for a uniform beam with equal spans and uniform loading throughout, derived an equation relating the reactions of any three consecutive supports (assuming all supports are rigid and at the same level). Rebhann was, incidentally, among the first to show bending moment diagrams for beams (Todhunter dismisses his work as insignificant, however). Köpeke (1856), Belanger (1858, 1862) and Scheffler (1857, 1858b) considered the effect of difference in level between the supports of continuous beams. Belanger was Clapeyron's pupil: he considered the problem of a beam with only two spans which, however, had different flexural rigidities and uniformly distributed loads. Scheffler's contribution includes an attempt to establish graphical analysis of uniformly loaded beams on several supports, by reference to points of contraflexure and maximum and minimum curvature. Then later (1860), he made an exhaustive analysis of a three-span beam for various conditions of loading, as well as of span ratios and support levels. But the most important developments after Clapeyron seem to be due mainly to Heppel, Mohr, Winkler, Bresse, Weyrauch and W. Ritter. Also, it is appropriate to acknowledge the exhaustive account of bridge engineering by Molinos & Pronnier (1857), with its discussion of experimental and theoretical investigations, having regard to results obtained by Hodgkinson and Fairbairn as well as by Clapeyron and Belanger.

Heppel, 1858

Heppel's theory of continuous beams is contained in two papers: the first, presented to the Institution of Civil Engineers (1858), deals with the analysis of beams whose spans have uniformly distributed and central concentrated loads. He derives a form of Clapeyron's theorem and, in addition, finds reactions of supports, points of maximum stress and contraflexure and deflexions. Heppel's second paper (1870), with its historical survey, is especially interesting, however.

Having remarked on Navier's original method he writes as follows:

'This method, although perfectly exact for the assumed conditions, was objectionable from the great labour and intricacy of the calculations it entailed'. Heppel remarks that Molinos & Pronnier explained the method fully and showed that for a bridge of n spans, $3n + 1$ equations are involved and quoted the example of a bridge of 6 spans requiring 19 simultaneous equations to find its bending moments. He continues by remarking that Navier's method was the only one available until about 1849 when Clapeyron

being charged with the construction of the Pont d'Asnières, a bridge of five continuous spans over the Seine, near Paris, applied himself to seek some more manageable process. He appears to have perceived (and so far as the writer is informed, to have been the first to perceive) that if the bending moments over the supports at the ends of any span were known, as well as the amount and distribution of the load, the entire mechanical condition of this portion of the beam would become known just as if it were an independent beam.

According to Heppel, Clapeyron found himself obliged to introduce additional unknowns ('inconnues auxiliaires') into his equations, namely the slopes of the deflexion curve at the points of support. He was, therefore, compelled to operate on a number of equations equal to twice the number of spans. Remarking that Clapeyron 'does not appear as yet, to have made any formal publication of his method', Heppel proceeds to credit Bertot with the theorem of three moments. It seems, therefore, that he was unaware of Clapeyron's discourse of 1857.

Heppel acknowledges Bresse for the next major contribution to the subject and he also mentions Belanger, Albaret, Collignon, Piarron de Mondesir and Renaudot, as well as Molinos & Pronnier, with regard to the development of analytical techniques in practice.

With regard to his own experience, Heppel acknowledges Moseley

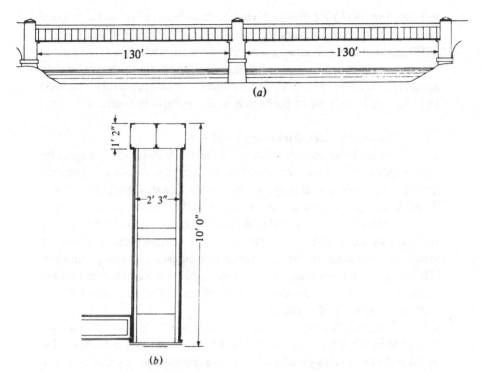

Fig. 7. (*a*) Torksey tubular girder bridge – elevation. (*b*) Torksey tubular girder bridge – section of a girder.

(1843) for introducing Navier's method to Britain, and then Pole for practical applications, especially to the Torksey Bridge in 1849 (Fig. 7) and the Britannia Bridge (where 'some new conditions' were treated successfully for the first time). Then, in 1858–9, 'being Chief Engineer of the Madras Railway', Heppel had occasion to investigate the conditions of a bridge of five continuous spans over the River Palar. After remarking that the only books available to him were those of Moseley and Clark he wrote that after many attempts and failures, the same idea occurred to him which appeared to have struck Clapeyron nine or ten years before, that if the bending moments over the supports were known, the whole conditions would become known. Following this clue, he was fortunate enough to succeed at once in eliminating the other unknown quantities which Clapeyron had been obliged to retain in his equations for many years after his original discovery of the method, and thus he was able to arrive at an equation identical with that which had been first published in France by Bertot. (Heppel's remarks confirm his ignorance of Clapeyron's discourse published in 1857.) He checked what he then believed to be his original discovery with Pole's results for the Britannia Bridge (in Clark's book, 1850) and in so doing obtained a generalised equation which he subsequently found to be identical to that of Belanger. Heppel concludes: 'It would appear, then, that the theory of this subject was independently advanced to about the same state of perfection in France and in England.'

Bresse, Winkler, Mohr and Weyrauch

In his treatment of continuous beams Bresse (1859) begins by acknowledging Clapeyron but emphasises that he does not use precisely the same method for obtaining a similar result. The third part of his *Cours de mécanique* (1865) includes an extensive study of continuous beams to include small differences in level of supports and various load distributions and span ratios, together with an exhaustive examination of the worst location of live load. In his construction of diagrams showing variation of bending moment at various points, as specified loads traversed continuous beam, he came close to the discovery of influence lines (a concept due to Winkler, as noted in Chapter 3).

Credit for priority in studying the most unfavourable position of a load on a continuous beam seems to be due to Winkler (1862), however, who was also the first to analyse a long beam on a continuous elastic foundation (1867). He was then concerned with railway construction. His book on theory of bridges (1872) includes an extensive treatment of continuous beams and theory of arches. It is, incidentally, important to note that Winkler and his contemporaries included continuous trussed girders as well as continuous plate girders in their studies.

Mohr's first article (1860) deals with modification of Clapeyron's theorem to allow for small differences in level of the supports of a continuous solid or trussed beam. Also, there is graphical representation of bending moments and shearing forces as well as a variety of specific examples. One example is concerned with optimum differences in level of supports, for various span ratios, with regard to maximum bending moment, due to specified loading. (Comparison of this method of controlling bending moments in continuous beams with that used for the Britannia Bridge seems relevant.) Two additional articles under the same title appeared in 1862 and 1868. The former included analysis of continuous beams of varying section and the error incurred if uniformity of section is assumed to simplify calculations. The latter represented a major contribution to the subject of beam theory in general.

Mohr begins his third article (1868) with the following interesting commentary:

Professor Culmann in his book *Graphische Statik* has set himself the task of using the new geometry for solving some problems in engineering, which are amenable to graphical treatment. The very interesting and useful results which this work presents, would, we are convinced, have already found general acceptance if the overly academic appertinances of the new geometry had not frightened many engineers from studying it. We believe that in many cases, and in particular for those of practical importance, the tools of the old geometry would have sufficed and we intend to illustrate this by several examples in future. In the present work we have attempted a graphical treatment of the elastic line theory, a problem which has hitherto defeated even the new geometry.

(The new geometry is viewed more favourably by Chalmers, as noted in Chapter 4.)

Mohr provides a comprehensive treatment of theory of beam systems with novel graphical interpretations and aids to computation, including graphical solution of continuous beams by properties of the elastic line (which Love (1892, 1927) describes in detail). Also, he introduces the concept of influence lines for deflexion of linearly elastic beams, thereby involving the reciprocal theorem independently of Maxwell (in his book (1906) on topics in technical mechanics, Mohr acknowledges Winkler's priority for the actual concept of influence lines). Other important features of the article are the property of bending moment diagrams known as 'fixed points', and theorems relating the elastic line and bending moment diagrams of beams. One theorem specifies that the change in slope of the elastic line between points whose abscissae are x_1 and x_2 is proportional to the area of the bending moment diagram between those points. Another specifies that the moment of that area about the origin $x = 0$ is proportional to the distance between the intercepts on an axis through the origin of the tangents to the elastic line at x_1 and x_2. Also, those tangents intersect below the centre of the area of that portion of the bending moment diagram if the beam is uniform. Otherwise, the diagram of M/EI must be used, EI being the flexural rigidity (Mohr used the symbol T instead of I).

Mohr also introduces the concept of elastic weights in the same article (a concept used implicitly by Culmann (1866), with reference to elastic arches), whereby deflexions of beams may be calculated by means of quasi-bending moments caused by distributed loading of intensity numerically equal to M/EI, where M is the bending moment caused by the actual loading of the beam (Fig. 8).

A comprehensive treatment of the theory of continuous beams is given

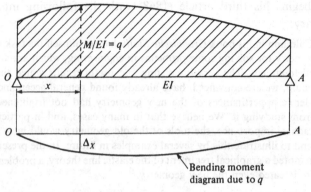

Fig. 8

by Weyrauch (1873), who provided perhaps the first exhaustive treatise on the subject, which achieved the status of an authoritative standard work for practising engineers. (As noted in Chapter 10, Weyrauch has been credited with the terminology 'influence line' to describe the well-known concept due to Winkler and developed in a variety of ways by Mohr and especially by Müller-Breslau.)

The subject of continuous beams is an important feature, moreover, in the second volume of Levy's celebrated book which appeared some twelve years (1886) after the first volume, and almost simultaneously with his important article on the elastic theory of suspension bridges (Chapter 3).

Graphical method: Fidler and Müller-Breslau

The history of theory of continuous beams in the nineteenth century would be less than adequate without an account of an important graphical method. Without seeking to belittle the graphical methods of others, notably Mohr (1868), attention is confined here to the method of characteristic points due to Fidler (1883), which was improved by Müller-Breslau in 1891 and included in his *Graphische Statik der Baukonstruktion* (1892, vol. 2, p. 357). Characteristic points for a uniform span of l are located at abscissae of $l/3$ and $2l/3$ and ordinates determined by:

$$\left. \begin{aligned} h_1 &= \frac{2}{l^2} \int_0^l M'(l-x)\mathrm{d}x \\ h_2 &= \frac{2}{l^2} \int_0^l M'x\,\mathrm{d}x \end{aligned} \right\} \tag{2.33}$$

and

where M' is the statically-determinate bending moment due to the load of the span. These quantities are shown in Fig. 9(b). If there are terminal

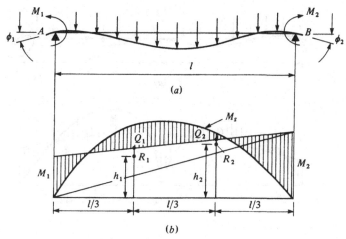

Fig. 9

couples applied to the span or beam, M_1 at $x = 0$ and M_2 at $x = l$, then if the supports are rigid at the same level it may be shown that (Salmon, 1938, vol. 1, p. 141):

$$\left.\begin{array}{l} \phi_1 l = \dfrac{l^2}{2EI} \left\{ \left(\dfrac{2M_1}{3} + \dfrac{M_2}{3} \right) - h_1 \right\} \\[3mm] \phi_2 l = \dfrac{l^2}{2EI} \left\{ \left(\dfrac{M_1}{3} + \dfrac{2M_2}{3} \right) - h_2 \right\} \end{array}\right\} \tag{2.34}$$

where ϕ_1 and ϕ_2 are the slopes due to bending of the beam at $x = 0$ and $x = l$ respectively (Fig. 9(a)). Noting that the first two terms in parentheses (in each of the above equations) represent the ordinates of the bending moment diagram due to the terminal couples, at $x = l/3$ and at $x = 2l/3$ respectively, and denoting them by p_1 and p_2 gives:

$$\left.\begin{array}{l} \phi_1 = \dfrac{l}{2EI}(p_1 - h_1) = \dfrac{l}{2EI} q_1 \\[3mm] \phi_2 = \dfrac{l}{2EI}(p_2 - h_2) = \dfrac{l}{2EI} q_2 \end{array}\right\} \tag{2.35}$$

where: $q_1 = p_1 - h_1$ and $q_2 = p_2 - h_2$. The derivations involved are conveniently accomplished by so-called moment–area methods to which Mohr contributed extensively but whose origins can be traced to Saint-Venant (1864).

For an encastré beam $\phi_1 = \phi_2 = 0$ and, therefore, the characteristic points lie on the straight line of the terminal couple bending moment diagram so that it is a simple matter to find those couples if the

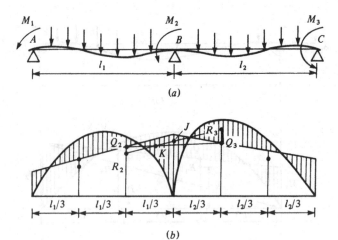

Fig. 10

characteristic points are known. Continuous beams present a more formidable problem and a graphical procedure by trial and error may be used. If two uniform adjacent spans are AB and BC (Fig. 10(a)), enjoying common conditions at B, and if the slopes are ϕ_A, ϕ_B and ϕ_C over the supports at A, B and C respectively, then:

$$
\left.
\begin{aligned}
\phi_A &= \frac{l_1}{2EI} q_1 \\
\phi_B &= \frac{l_1}{2EI} q_2 = -\frac{l_2}{2EI} q_3 \\
\phi_C &= \frac{l_2}{2EI} q_4
\end{aligned}
\right\}
\tag{2.36}
$$

where: $AB = l_1$; $BC = l_2$; and q_1 and q_2 refer to AB; and q_3 and q_4 refer to BC. Thus ϕ_B, being common to both spans, affords a relationship which is the basis of Fidler's method, that is:

$$
\left.
\begin{aligned}
& l_1 q_2 = -l_2 q_3 \\
\text{or} \quad & l_1(p_2 - h_2) = -l_2(p_3 - h_3)
\end{aligned}
\right\}
\tag{2.37}
$$

when EI is constant.

If now, as shown in Fig. 10(b), a straight line is drawn joining the characteristic points R_2 and R_3 and if $R_2 J = R_3 K$ (where K is chosen on that basis), then by using the result derived above it may be shown that a straight line joining the points Q_2 and Q_3 passes through K which is therefore known as the 'intersection point'. The graphical process consists essentially of finding the bending moments at A, B and C, which meet this condition as being the statically-indeterminate quantities required of the solution. It is a trial and error process which can be achieved rapidly with the aid of a tensioned thread and pins (see Salmon, 1938, vol. 1, p. 144). In common with the analytical methods, the method of characteristic points may be developed to take account of differences in level of supports and non-uniformity.

Soon after Fidler published his graphical method for analysing continuous beams, W. Ritter (1886) published an extension of Mohr's use of the elastic line to simplify continuous beam analysis.

Clebsch

While railway engineers developed theory of continuous beams, a significant treatment of the problem was produced by the mathematician Clebsch (1862) within his researches into theory of elasticity. It does not seem, however, to have become either well known or advantageous in engineering but it bears some similarity to Scheffler's method (1857; 1858 c).

Clebsch's elegant use of the differential equation of bending seemed to be unknown in Britain until the method was published in 1919 independently, it is believed, by Macaulay and subsequently termed 'Macaulay's method'. A few years before the close of the century, Wilson's so-called method (1897) for continuous beam achieved popularity in Britain. It is relevant to Clebsch's method, being concerned with obtaining equations relating loading and reactions of intermediate, simple supports (redundants) by determining the conditions for no deflexion at those supports, when the beam is regarded as supported at its extremities with forces representing loads and intermediate reactions applied to it. Clebsch's ingenious method of integration and use of boundary conditions contrasts with Wilson's physical reasoning.

In article 87 of the translation of Clebsch by Saint-Venant and Flamant (which has no illustrations), a uniform beam on numerous point supports is first considered with the beam simply supported with loading, including point loads P_1, P_2, ..., P_n, at intervals l_1, l_2, ..., l_n along the span corresponding to the positions of the intermediate supports. The flexural rigidity is denoted by $E\sigma\lambda^2$; the bending moment due to the loads other than P_1, P_2, ..., P_n, by M; the deflexion by u; abscissae by z; and constants of integration by α and β.

After thus dealing with the problem of a simply supported uniform beam with n point loads, Clebsch proceeds to use the same procedure for dealing with the problem posed by intermediate, simple supports, that is, the problem of the continuous beam. He simply treats their reactions as upward loads and obtains sufficient equations for evaluating them by using

34   *Beam systems*

the conditions for zero deflexion of those points. For a uniformly distributed load of intensity Π with $(n-1)$ equally spaced intermediate supports, distant l apart, he obtains simultaneous equations relating the reactions $Q_1, Q_2, \ldots, Q_{n-1}$ of those supports, as follows:

$$\left.\begin{array}{c} Q_1+4Q_2+Q_3 = 6\Pi l \\ Q_2+4Q_3+Q_4 = 6\Pi l \\ Q_3+4Q_4+Q_5 = 6\Pi l \\ \vdots \end{array}\right\} \qquad [(2.38)]$$

The similarity of form of these equations and those of Clapeyron's theorem of the three moments is remarkable.

It is regrettable that Clebsch's method was introduced in the realms of the mathematician and remained unknown to engineers generally, for many years. There is no doubt, however, of its status with regard to the history of the subject.

Notes

Todhunter (1892) presents a confused impression of the principles used to optimise the dead-load bending moment distribution of the Britannia Bridge. It is asserted that the objective was to equalise the 'pressures' on the supports rather than bending moments and also that Clapeyron's (1857) criticism is inappropriate because the bridge 'is not a continuous beam in the theoretical sense'. Moreover, it is apparently suggested that Clapeyron's theorem could have been used to advantage for the design of the bridge, whereas that process ante-dated the theorem by several years. (Todhunter also criticises Navier because induced axial tension is omitted from analysis of an encastré beam.)

Jourawski criticised (1856) the Britannia Bridge for lack of shear strength at the piers with regard to rivet pitch at joints, in the light of his then novel theory of shear stress distribution in beams (a similar theory was proposed by Rankine some two years later, in 1858, and Timoshenko (1953, p. 89) refers to unpublished work on shear stress by Poncelet).

Bell (1872) describes his difficulty in applying Navier's theory to the continuous girders of the Chepstow Bridge by Brunel in 1849 but records that the results of his calculations for continuity over five spans were confirmed by experiment. (He applauds the discovery of the theorem of the three moments with reference to Heppel (1870).)

Wilson, whose paper on analysis of continuous beams (1897) was communicated to the Royal Society by Osborne Reynolds, was at that time a demonstrator at Owen's College, Manchester. At the beginning of the paper he acknowledges its basis as a principle due to Bresse, which is now known as the principle of superposition (see also Chapter 10: Notes).

Saint-Venant refers, on p. 104 of his edition (1864) of Navier's *Leçons*, to an error in theory of flexure in Gauthey's treatise on bridges (1813, edited by Navier). The error was rectified in the lithographed notes for Navier's course (1819–20) at l'Ecole des Ponts et Chaussées and is said to be traceable to James Bernouilli and Mariotte.

3

Theory of the arch and suspension bridge

At the beginning of the nineteenth century, to which elastic arch theory belongs, masonry was still the principal structural material. Many major structures, especially bridges, depended on the arch as a means of exploiting the strength of stone in compression. The origin of an explicit theory of the arch is variously ascribed to Hooke, De La Hire, Parent and David Gregory in the seventeenth century. Robison (Brewster, 1822) believed that Hooke suggested the inversion of the shape adopted by a suspended rope or chain, namely the catenary, as the statically correct form for an arch: others (Straub, 1952) ascribed that concept to David Gregory. In any event, it appeared to disregard a distribution of load different from that which would be due to a uniform voussoir arch. Heyman has reviewed the development of the theory of the arch in detail (1972) and leaves little doubt that it was highly developed in the eighteenth century. Coulomb's theory of 1773 (1776) of the distribution of force in loaded stone arches and their stability (ultimate load carrying capacity) was generally accepted by Navier, to judge by the contents of his *Leçons* (1826, 1833; in which, quite separately, elastic theory of the arch rib appears). But those youthful partners, Lamé & Clapeyron rediscovered (while in Russia) the theory of the ultimate strength of stone arches for themselves (1823), apparently ignorant of Coulomb's theory (or, indeed, that of Couplet which Heyman has described (1972)). For practical purposes, since arches carried heavy dead load, due to superstructure, in comparison with which live load was small, elementary statics was sufficient to ensure that the distribution of dead load and arch shape were such that the locus of the resultant of the total shear force at any section and the horizontal thrust, nowhere passed outside the masonry.

Masonry arch

In Britain the process of rediscovery of the theory of the masonry arch began somewhat later and seems to have been due to the Rev. Professor Henry Moseley. Thus, in 1835, after an interval of two years since it was presented, Moseley's paper on the equilibrium of bodies in contact was published and he observed finally that 'the great arches of late years erected by Mr Rennie in this country have for the most part been so loaded as very nearly to satisfy his condition for stability'. That condition specified that in the absence of friction, the pressure between voussoirs should be transmitted such that the direction in which it acts at each joint is at right angles to the surface of contact. Then, in the preface of his book (1843), Moseley, in referring to his 'memoir' on the stability of bodies in contact (1835) as the basis for his theory of the arch, remarks that his principles differ essentially from those on which the theory of Coulomb is founded but, nevertheless, when applied to similar problems, they give identical results. He also notes that the theory of Coulomb was unknown to him at the time his memoir was published and refers to Hann's treatise (1843) for a comparison of the two theories. Thus it seems that there is agreement with Coulomb (1776) that, for stability of an arch, the line of thrust must not pass outside the masonry at any point. On the basis of his 'new principle in statics' (the principle of least resistance or pressures (1833*a*) he proposed, in addition, that among all lines of thrust for an arch, which satisfy the primary condition of stability, that which is consistent with minimum thrust at the crown is correct. (He had, moreover, illustrated his principle with reference to Euler's problem of a solid resting on four supports, which is noted in Chapter 7.)

Engineers, including Barlow (1846) and Snell (1846), showed interest in Moseley's theory, as did Scheffler (1858*c*) in Europe, who also translated his book in 1844. There seemed a total lack of awareness, however, which was to persist to an extent for many years, that the thrust-line (and therefore the horizontal thrust) for the dominant dead load of a masonry arch is not unique: this is due, in part at least, to the inevitable lack of homogeneity of masonry. But stability is assured if it is *possible* to construct a thrust-line which does not pass outside the masonry at any point (Heyman, 1966).

Application of theory of elasticity to masonry arches belongs to the nineteenth century. It was suggested, notably by Poncelet (1852) in an extensive critical review of arch theory, which included Navier's theory of the elastic rib (1826) and Mèry's (*c.* 1840) unsatisfactory version, as well as Moseley's work continued by Scheffler. Villarceau (1853) also considered this approach but Heyman (1966) emphasises his ultimate firm rejection of it in favour of elementary statics as the basis of *his* rules for the design of masonry arch bridges which, according to Heyman, have never been superceded. Later in the nineteenth century, the emergence (ascribed to Navier) of the so-called 'middle third rule' for masonry structures was a consequence of appeal to theory of elasticity for homogenous materials. The rule is described in some detail by Salmon (1938) who includes Fuller's graphical analysis of masonry arches (1874), which was based upon it.

Brunel devised a novel numerical method of determining a thrust-line by elementary statics (Fig. 11), which Owen has described (Pugsley, 1976)

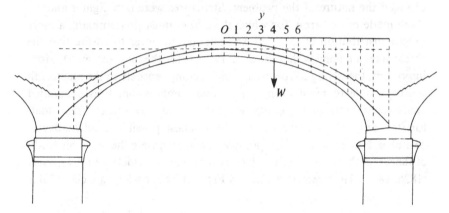

Fig. 11. After Brunel (*c*. 1840).

and which illustrates a leading practitioner's approach to arch theory. Taking the origin O at the crown he represented the dead load on the section between O and y, that is, the shearing force W, by the polynomial:

$$W = Ay + By^2 + Cy^3 + Dy^4 \qquad [(3.1)]$$

where y is the horizontal distance from the crown and the parameters A, B, C, D are determined by the values of W calculated for four chosen points within a half-span using the assumed shape of the arch. Expressing W in terms of the horizontal component H of the thrust he wrote:

$$W = H\,dx/dy = Ay + By^2 + Cy^3 + Dy^4 \qquad [(3.2)]$$

and integrated to obtain:

$$Hx = Ay^2/2 + By^3/3 + Cy^4/4 + Dy^5/5 + K \qquad [(3.3)]$$

as the equation of the thrust-line. The constant of integration K is zero since $x = 0$ when $y = 0$. Now the value of H may be found by choosing a point (y, x) through which the thrust-line is required to pass and which complies with the distribution of shearing force. A safe design is represented by an arch whose centre line is specified by the equation for *x* with the calculated value of *H*. The fact that the value of *H* is not unique is immaterial so long as there is a thrust-line which is accommodated within the thickness of the arch. Brunel's calculations for several arches are preserved and Owen (Pugsley, 1976) has given those of the Maidenhead Bridge.

Elastic arch: Navier, 1826

The development of the metal arch in the nineteenth century changed the nature of the problem. Structures were now lighter and for those made of cast iron, for which dead load must predominate, a more precise distribution of the load was essential in order to ensure that the thrust-line remained within the arch itself (as for the masonry arch). More important, the wrought iron arch (which belongs entirely to the nineteenth century), able to resist both tension and compression, thus demanded analysis within the theory of elasticity. It was not dependent on dead load for its stability; on the contrary, the smallest possible dead load was desirable. Navier had, in fact, provided an appropriate theory in his study of the strength of elastic ribs, which is contained in article 6 of his *Leçons* (1826, 1833). He begins (Fig. 12; his Fig. 72) by considering a curved bar

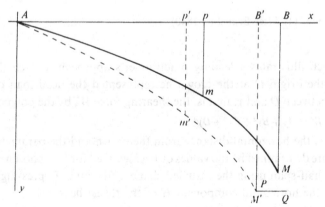

Fig. 12. From Navier (1826).

encastré at its left-hand end *A* and loaded at its free end *M* by forces *P* and *Q* vertically and horizontally. The equilibrium of the rib is expressed by the equation:

$$\epsilon \frac{\mathrm{d}\phi' - \mathrm{d}\phi}{\mathrm{d}s} = P(a-x) + Q(b-y) \qquad\qquad [(3.4)]$$

where ϵ is the flexural rigidity and $\mathrm{d}\phi' - \mathrm{d}\phi$ is the change of the angle of curvature (related to the vertical axis) of an element of length $\mathrm{d}s$ due to flexure; a and b are the coordinates of M. Therefore:

$$\phi' - \phi = \frac{1}{\epsilon} \int \mathrm{d}x \sqrt{1 + \left(\frac{\mathrm{d}y}{\mathrm{d}x}\right)^2} [P(a-x) + Q(b-y)] \qquad [(3.5)]$$

Navier continues in an unfamiliar manner which is not conducive to general formulae but rather to providing results for specific shapes. Thus, having noted that he is concerned with bending, whereby the difference between ϕ' and ϕ is small, he writes:

$$\left.\begin{aligned}
\cos \phi' - \cos \phi &= -\frac{1}{\epsilon} \sin \phi \int \mathrm{d}x \sqrt{1 + \left(\frac{\mathrm{d}y}{\mathrm{d}x}\right)^2} [P(a-x) + Q(b-y)] \\
\sin \phi' - \sin \phi &= \frac{1}{\epsilon} \cos \phi \int \mathrm{d}x \sqrt{1 + \left(\frac{\mathrm{d}y}{\mathrm{d}x}\right)^2} [P(a-x) + Q(b-y)]
\end{aligned}\right\} \qquad [(3.6)]$$

then since $\cos \phi = \mathrm{d}x/\mathrm{d}s$, $\sin \phi = \mathrm{d}y/\mathrm{d}s\dots$:

$$\left.\begin{aligned}
\mathrm{d}x' - \mathrm{d}x &= -\frac{1}{\epsilon} \mathrm{d}y \int \mathrm{d}x \sqrt{1 + \left(\frac{\mathrm{d}y}{\mathrm{d}x}\right)^2} [P(a-x) + Q(b-y)] \\
\mathrm{d}y' - \mathrm{d}y &= \frac{1}{\epsilon} \mathrm{d}x \int \mathrm{d}x \sqrt{1 + \left(\frac{\mathrm{d}y}{\mathrm{d}x}\right)^2} [P(a-x) + Q(b-y)]
\end{aligned}\right\} \qquad [(3.7)]$$

equations which, he says, may be integrated to give the displacements of any point of the rib if its shape is specified. It is interesting to note Navier's apparent lack of consistency with regard to signs of applied forces, various instances of which occur in his *Leçons*, but do not detract materially from the clarity of his expositions. Thus, in specifying x positive to the right and y positive downward for the example of a downward curving rib, and having indicated positive senses of P and Q accordingly, he gives the bending moment at (x, y) as $P(a-x) + Q(b-y)$, instead of $P(a-x) - Q(b-y)$ or, for the alternative convention, $-P(a-x) + Q(b-y)$. The former appears to be appropriate because the rib is shown as having increased curvature due to the applied forces. Also, there is no system of numbering or identifying equations in Navier's work.

The first example given by Navier is a rib of parabolic shape for which $y = bx^2/a^2$; $\mathrm{d}y/\mathrm{d}x = 2bx/a^2$. Substituting in the simplified equations and replacing the square root by a series:

$$\left.\begin{aligned}
\mathrm{d}x' - \mathrm{d}x &= -\frac{1}{\epsilon} \mathrm{d}y \int \mathrm{d}x \left[1 + \frac{1}{2}\left(\frac{\mathrm{d}y}{\mathrm{d}x}\right)^2 + \dots\right] [P(a-x) + Q(b-y)] \\
\mathrm{d}y' - \mathrm{d}y &= \frac{1}{\epsilon} \mathrm{d}x \int \mathrm{d}x \left[1 + \frac{1}{2}\left(\frac{\mathrm{d}y}{\mathrm{d}x}\right)^2 + \dots\right] [P(a-x) + Q(b-y)]
\end{aligned}\right\} \qquad [(3.8)]$$

he obtains for the horizontal and vertical displacements (h and f) due to flexure:

$$-h = \frac{P}{\epsilon}\left(\frac{5a^2b}{12} + \frac{b^3}{10} + \ldots\right) + \frac{Q}{\epsilon}\left(\frac{8ab^2}{15} + \frac{16b^4}{105a} + \ldots\right) \Big\}$$

$$f = \frac{P}{\epsilon}\left(\frac{a^3}{3} + \frac{ab^2}{15} + \ldots\right) + \frac{Q}{\epsilon}\left(\frac{5a^2b}{12} + \frac{b^3}{10} + \ldots\right) \Big\}$$

[(3.9)]

Next he considers such a rib with its ends resting on a smooth surface with a concentrated load 2Π applied to its mid point or crown. He notes that now $P = -\Pi$ and $Q = 0$ whence:

$$h = \frac{\Pi}{\epsilon}\left(\frac{5a^2b}{12} + \frac{b^3}{10}\right) \Big\}$$

$$-f = \frac{\Pi}{\epsilon}\left(\frac{a^3}{3} + \frac{ab^2}{15}\right) \Big\}$$

[(3.10)]

the latter representing the vertical deflexion at the load.

Again he considers Fig. 13 (Fig. 74 of *Leçons*), a parabolic rib with a central vertical load 2Π supported at its ends on the same horizontal level

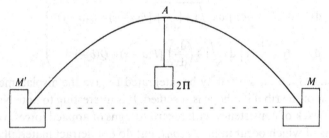

Fig. 13. From Navier (1826).

in such a manner that horizontal movement cannot occur (as in a two-pin elastic arch). By the condition that $h = 0$ and noting that $P = -\Pi$, he finds that:

$$Q = \Pi\left(\frac{25a}{32b} - \frac{b}{28a}\right) \Big\}$$

$$f = -\frac{\Pi}{\epsilon}\left(\frac{a^3}{128} - \frac{23ab^2}{6720}\right) \Big\}$$

[(3.11)]

Apart from considering distributed loading and ribs of circular shape Navier does not develop the topic further and proceeds to discuss the nature of its applications. Subsequently, in article 10 about timber bridges, there is a section on 'bridges supported by arches' (that is, those which rely on timber lattice arch ribs as shown in Figs. 14 and 15 (his Figs. 138 and 139)). He uses the principle of the two-pin elastic arch for their analysis and shows how the thrust within the arch may be found (and in so doing he comes close to a similar approach for masonry arches). Then he deals

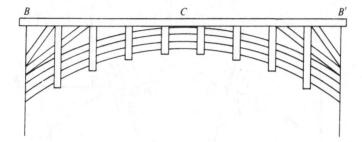

Fig. 14. From Navier (1826).

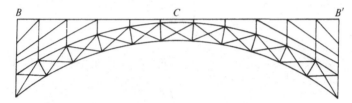

Fig. 15. From Navier (1826).

with the topic of timber centring for the construction of voussoir arches and the forces which it must resist. In footnotes he refers to Rennie, and to Robison's section on carpentry in his *System of mechanical philosophy* (Brewster, 1822). There is, moreover, a footnote (on p. 406) with reference to the cast iron Pont d'Austerlitz in Paris and the similar bridge at Sunderland, in relation to similarity to timber construction. (It is a feature of the earliest iron bridges that in detail they closely resembled timberwork which provided the only relevant precedent with regard to technique.)

Bresse, 1854

It seems that after Navier it was Bresse to whom further elucidation and progress regarding the theory of curved elastic bars is due. In 1854 he published *Réchérches analytiques sur la flexion et la résistance des pièces courbes*, having in the previous year succeeded Belanger as professor of applied mechanics at l'Ecole des Ponts et Chaussées. He addressed himself to finding, for an arch rib, the stresses due to specified loads and reactions; the effects of temperature changes on stress and deflexion; and, for specified loads and end constraints, the reactions incurred. (Todhunter & Pearson, (1893, vol. 2(i), p. 352) remark on the extensive tables given by Bresse to facilitate calculations of the behaviour of ribs of circular form. Villarceau (1853) is similarly commended for his provision of tables

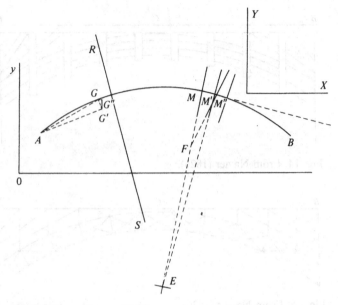

Fig. 16. From Bresse (1854).

relating to his differential equation for the 'pressure line' of a rigid arch to ensure coincidence in the shapes of both.)

Bresse derived equations of the following form for an arched rib, using the single, somewhat obscure, diagram shown in Fig. 16:

$$\Delta\alpha - \Delta\alpha_0 = \sum_{S_0}^{S} \frac{M}{\epsilon} \delta s \qquad [(3.12)]$$

with regard to angle of bending from an origin S_0 along the rib at (x_0, y_0), where the angle of the tangent is specified by α_0; M is the bending moment at any point; and ϵ is the flexural rigidity:

$$U - u_0 = \Delta\alpha_0(Y - y_0) + \beta(X - x_0) + \sum_{S_0}^{S}\left[(Y - y)\frac{M}{\epsilon} + \frac{N\,\mathrm{d}x}{\epsilon\,\mathrm{d}s}\right]\delta s \qquad [(3.13)]$$

for the horizontal deflexion relative to that of the chosen origin of a point (X, Y) distant S along the rib, and:

$$V - v_0 = -\Delta\alpha_0(X - x_0) + \beta(Y - y_0) + \sum_{S_0}^{S}\left[-(X - x)\frac{M}{\epsilon} + \frac{N\,\mathrm{d}y}{\epsilon\,\mathrm{d}s}\right]\delta s \qquad [(3.14)]$$

for the vertical deflexion, where β is a coefficient of elasticity and s is the distance of any point (x, y) between S_0 and S.

In the last two equations the first term of each refers to deflexion due solely to rotation of the chosen origin; the second to stretch of the rib due to causes other than loading, for example temperature change; the third to deformation by the loading manifest by axial force N and bending moment M at any point (X, Y). Displacements are assumed to be small and consistent with the assumption of constant geometry of an arch.

Bresse gives practical applications including those to a uniformly loaded, uniform rib of circular form and, of special interest, to the circular rib of a cast iron railway bridge at Tarascon over the Rhône, where measurements were taken on the rib before and after erection, it seems. Bresse's value of deflexion by calculation was apparently only 0.0008 m less than the measured value of 0.0650 m.

The possible circumstances of statical indeterminacy are examined with regard to terminal conditions, and their use in the deflexion equations are described for the purpose of achieving a solution. He does not confine attention to single arches but deals as well with combinations of ribs and the conditions for compatibility of displacements at junctions (necessary to the study of their resistance to loading and temperature variations). The multiple arch iron bridge at Vergnais is used to illustrate these aspects.

Principle of superposition: symmetry and anti-symmetry

The final part of Bresse's third chapter 'Remarques et théorèmes concernant la manière dont les forces extérieures entrent dans les formules de la flexion' is of special interest and importance for its exploitation of the principle of superposition to facilitate analysis where multiple loads are involved, and particularly for the powerful device of symmetry and anti-symmetry which follows therefrom. Thus, if an arch is symmetrical (in respect of both shape and elasticity) about, say, a vertical axis, and it carries a concentrated load Π acting parallel to that axis (Fig. 17) thereby causing horizontal reactions at its extremities Q_1 and $-Q_1$ (since, in the absence of other than vertical loading there cannot be a resultant horizontal force applied to the arch), then if the load is reversed to become $-\Pi$ it will cause reactions $-Q_1$ and Q_1 respectively. Again, equal loads of Π applied symmetrically with regard to the vertical axis will clearly cause horizontal reactions $Q_2 = Q_1 + Q_1 = 2Q_1$ and $-Q_2 = -Q_1 - Q_1 = -2Q_1$ respectively (viewing from back and front of the arch verifies this), while equal and opposite loads so applied will cause reactions $Q_1 - Q_1 = 0$ and $-Q_1 + Q_1 = 0$ respectively. It follows therefore that a single force of 2Π will cause horizontal reactions of $2Q_1$ and $-2Q_1$, precisely as if two loads Π were applied symmetrically. The principle of superposition and its related device of symmetrical and anti-symmetrical components became powerful aids, together or individually, for the analysis of every kind of structure with linear characteristics.

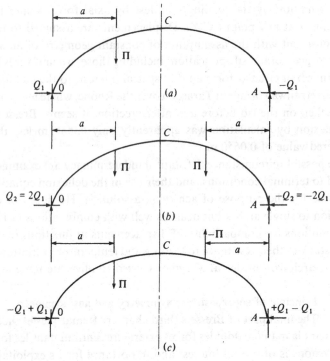

Fig. 17

The remainder of Bresse's work on arches (chapters 4 and 5) is concerned with what are, in the present context, matters of detailed calculation. For example, the final chapter investigates the most severe effects of distributed loading of a railway arch and optimum arch sections in relation to the ratio of rise to span for circular arches (on which Bresse concentrates throughout his calculations). His wider studies in applied mechanics embraced fluid mechanics (Rouse & Ince, 1957).

The elastic centre

The concept of the elastic centre of a solid curved elastic bar (the elastic arch) is probably due, in the first place, to Culmann (1866). Chalmers (1881) reproduces Culmann's analysis of an elastic arch, including Fig. 200 (Fig. 18) of his celebrated book. The equations for the deflexion, due to bending of the free end of an arch, encastré at the opposite end and loaded such that there is a bending moment M, at any element of length ds and flexural rigidity EI, whose location is defined by coordinates y, z with reference to the free end, are given as follows:

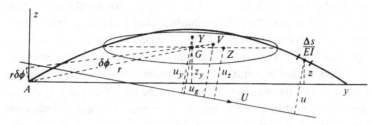

Fig. 18. After Culmann (1866).

$$\left.\begin{aligned}
\phi &= \Sigma\, M \frac{ds}{EI} \\
\Delta y &= \Sigma\, z\, M \frac{ds}{EI} \\
\Delta z &= \Sigma\, y\, M \frac{ds}{EI}
\end{aligned}\right\} \qquad\qquad [(3.15)]$$

If now U is the resultant of the forces acting on the arch and causing bending moments M, and the perpendicular distance from its line of action to an element ds is u; while that to the centre of gravity G of the arch with respect to $\Sigma\, ds/EI = S$ is u_g, then:

$$\phi = U\Sigma u\, ds/EI = Uu_g S \qquad\qquad [(3.16)]$$

Thus the resultant of the forces is equivalent to a couple $Uu_g = M_g$, together with a force U through the elastic centre G, and the total angle of elastic bending ϕ depends only on that couple.

It may further be deduced from Culmann's concept that the displacements of the orthogonal components of the force U, with reference to the elastic centre, may be identified independently. If the arch is completely symmetrical so that its principal axes, with respect to $\Sigma\, ds/EI$ through G, are parallel to the axis of y and z, then those components of elastic displacement are:

$$\left.\begin{aligned}
\Delta_y' &= U_y \Sigma (z-z_g)^2 \frac{ds}{EI} = \Delta_y + \phi y_g \\
\Delta_z' &= U_z \Sigma (y-y_g)^2 \frac{ds}{EI} = \Delta_z - \phi z_g
\end{aligned}\right\} \qquad\qquad [(3.17)]$$

U_y and U_z being the components of U in the y and z directions respectively; and y_g and z_g the coordinates of G. Hence $\Delta'y$ depends only on U_y and $\Delta z'$ only on U_z.

The use of the elastic centre as the point of reference for the resultant force on an arch in the manner described, together with the principal axes of 'elastic weight' ($\Sigma \mathrm{d}s/EI$), affords a very convenient means of analysing symmetrical statically-indeterminate arches and similar structures. For an encastré arch the resultant of the reactions at one abutment is equal and opposite to the resultant of the other forces acting on the arch. Since the deflexion, including rotation, at the abutments is zero, the deflexion of the resultant and its reaction (referred to the elastic centre) is also zero and the three equations which specify that condition are independent, each containing one component of the three which together comprise the resultant or total reaction.

The benefits conferred by the elastic centre were sought for framed structures and arches by Mohr and Müller-Breslau, as noted in Chapter 10; the underlying general concept being to choose the statically-indeterminate variables in such a manner that the relevant equations of compatibility of deflexion (strain) are simple ('normal'). Each equation then contains only one unknown.

Elastic analysis of rigid arches

Winkler (1867) and Mohr (1870) seem to have followed Poncelet (1852) in proposing theory of elasticity for the analysis of masonry arches as well as metal arches. In the light of subsequent research, notably by Heyman (1966), it seems as though they were mistaken, however (in spite of the popularity of the theory of elasticity of masonry arches ever since). But an approach whereby a masonry arch is regarded as a two-pin arch and its shape determined by successive approximation until it is identical with the bending moment due to the dead load, would appear to be valid (Fig. 19). It would, moreover, meet Winkler's *extremum* principle (1879*a*)

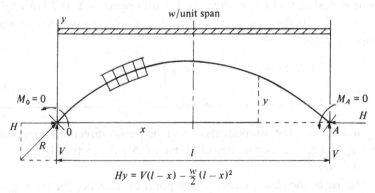

$$Hy = V(l - x) - \frac{w}{2}(l - x)^2$$

Fig. 19. After Navier (1826).

which is discussed below, since then the thrust-line would coincide with the centre line of the arch, provided that live loads are small in comparison with dead load.

In Britain, Bell, sometime assistant to Brunel, seemed abreast with developments in theory of arches on the continent of Europe when he presented his paper on stresses of rigid arches, continuous beams and curved structures to the Institution of Civil Engineers (1872). He derived the equations for the terminal deflexions of an elastic arch rib, due to bending, in accordance with the principles used by Navier and Bresse. Then he sought to simplify arch analysis for the engineer by graphical methods based on statics alone, and by the use of successive approximation to determine thrust-lines, such that the bending moments for a uniform encastré or rigid arch approximately satisfied the abutment conditions:

$$\left.\begin{array}{ll} \text{for slope} & \sum_0^l M\,\delta s = 0 \\[2mm] \text{for horizontal deflexion} & \sum_0^l My\,\delta s = 0 \\[2mm] \text{for vertical deflexion} & \sum_0^l Mx\,\delta s = 0 \end{array}\right\} \qquad [(3.18)]$$

The coordinates x, y, of an element of arch δs long, are measured from an origin at an abutment and the length l is the total length round the curve of the arch. He was, in fact, advocating theory of elasticity for masonry as well as iron arches. Thus, Professor E. Collignon said in the 'Discussion' that in applying the method to the masonry arch of the Pont-y-tu-Prydd: 'The great lightness of this arch, by rendering it sensible to the influence of accidental loading, made it a natural introduction to the study of metallic arches in which the action of external loads preponderated.' Clerk Maxwell and Fleeming Jenkin also contributed to the 'Discussion' which ranged widely within the scope of the title of the paper.

Mohr and Winkler: influence lines

In 1870 Mohr proposed an elegant graphical device of arch analysis. With reference to the two-pin arch shown in Fig. 20, it is noted that in the absence of horizontal restraint at C the bending of an element of length δs at (x, y) causes a horizontal displacement at the abutment of $My\,\delta s/EI$. If the bending is due to unit load at any distance $x = q$ from the origin at A, then the total horizontal displacement of the abutment would be:

$$h = \int^l \frac{My\,\mathrm{d}s}{EI} = \int^q \frac{x(l-q)y\,\mathrm{d}s}{lEI} + \int_q^l \frac{q(l-x)y\,\mathrm{d}s}{lEI} \qquad (3.19)$$

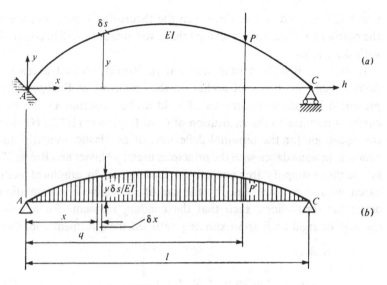

Fig. 20. After Mohr (1870).

which is of precisely the same form as that of the bending moment at a distance q along a simply supported beam of span AC with distributed loading whose intensity at any element x from A is $y\delta s/EI$. Since the magnitude of the horizontal thrust at the abutment to prevent horizontal displacement, due to the unit load, is directly proportional to the value of h, Mohr concluded that the bending moment diagram of a simply supported beam, due to distributed loading, whose intensity varies as $y\,\delta s/EI$, represents the influence line for abutment thrust of the arch. This is because the expression for h is the same for any value of q, or position of unit load on the arch.

Winkler introduced the concept of influence lines, initially for arches, as early as 1868 but the terminology 'influence line' is believed to be due to Weyrauch who made extensive studies of the theory of elastic arches in relation to bridge design (1878, 1896). Then, in 1879, Winkler (1879a) proposed the principle which bears his name and specifies that the correct line of thrust for an elastic arch is that which being statically admissible, is closest to the centre line of the arch. Being, at the time, evidently unaware of valid energy principles, including Castigliano's work (which he discovered after 1880), the precise justification for Winkler's theory is unclear. It is probable that its origin is belief in the concept of economy in Nature and the theorem of minimum energy for flexure, by D. Bernoulli and Euler, which Müller-Breslau (1886b) noted (Chapter 10). Fränkel (1882) gives details of Winkler's treatment, as described in Chapter 9.

Suspension bridges

The availability of wrought iron, strong in tension, toward the end of the eighteenth century, also led to the development of the suspension bridge as the inversion of the arch bridge. The suspension chains usually consisted of wrought iron bars pin-jointed together, with the deck supported from them by vertical rods. Deck structures were originally assumed to make no contribution to the strength of the structure and were so flexible that the chains withstood all the live load as well as the dead load of the structure itself. Those early suspension bridges were, in fact, very flexible and notorious for excessive oscillation, due to live loads and cross-winds. Moseley (1843) was aware of his phenomenon (Chapter 11): he refers to the collapse of Broughton Bridge near Manchester, which was due to the measured tread of soldiers; and he also mentions Navier's remarks on the problem. Navier's treatise on suspension bridges (1823) included the theory of the economical design of suspension chains, which was based on the elementary statics of a loaded hanging chain. This represented the theory of suspension bridges (as also exemplified by Lamé & Clapeyron (1826a)) until Rankine's theory of 1858. Thus, Telford's Menai Bridge (completed in 1826), and Brunel's design of 1829 for the Clifton Suspension Bridge, depended upon that theory, as Brunel's calculations clearly indicate (Charlton: Pugsley, 1976). Navier visited Telford's bridge during construction (while in Britain to study metal bridge construction on behalf of the French government) which was completed in the same year that his own suspension bridge over the Seine in Paris was abandoned (Appendix I).

The contribution of the gravity stiffness of a loaded chain or cable as the most significant factor in suspension bridge behaviour was not the subject of precise analysis until 1888. The effect was, however, observed by the middle of the century, for example by Roebling in 1855 at the time of the construction of his highly stiffened suspension bridge for carrying a railway over the Niagara (Pugsley, 1957). Also, the concept is suggested in an anonymous article (1860) within a detailed discussion of suspension bridge behaviour, including comparison with other kinds of bridge. Results of experiments on models are given, and in examining the results the author states: 'The other element of difference lies in the resistance of the chain itself to a change of position, a resistance very noticeable in a heavily weighted model. Verification is provided by quoting a one-third reduction in deflexion due to a concentrated load at a one-quarter-span point of a model if a distributed dead load, of rather more than three times the concentrated load, is present. Roebling's Niagara Bridge seemed to have provoked the article (and various others in favour of such bridges), for that bridge was the subject of introductory remarks which seemed to be directed at emphasising how untypical of suspension bridges it was, due to its very stiff girders and restraining chains. (It also stimulated letters from Rankine on suspension bridge theory.) Another anonymous article on suspension bridges appeared nearly two years later (1862), in which there is an attempt to quantify the gravity effect analytically. It seems likely that it was by the same author because the original article contained analysis of deflexion of a chain of constant length.

The first attempt at precise analysis of the composite behaviour of chain or cable system and deck with stiffening girders seems, then, to be due to Rankine (1858). His theory is well known and is part of elementary courses in theory of structures. It is based on the assumptions that the chain or cable is parabolic due to uniformly distributed dead loading of the deck structure, that the deck girders are unstressed in the absence of live load, and that live load is distributed in its entirety (but uniformly, regardless of its precise nature) by the deck girders to the cable through vertical suspension bars. Thus, if the deck girders are continuous, the live load intensity on the cable is simply the magnitude of that load, whenever situated on the deck, divided by the span. Thus the deck sustains the actual live load, together with reactions at each end of the span and the uniformly distributed reaction from the cable system, as shown in Fig. 21. If the deck

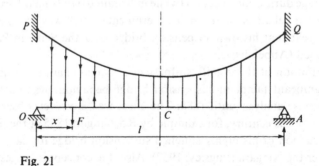

Fig. 21

girders have a central hinge, however, the magnitude of the live load, as uniformly distributed, may be determined by statics. Claxton Fidler followed Rankine in contributing to the theory of suspension bridges (1878) with reference to means of stiffening.

The elastic theory of suspension bridges

The effect of the relative elasticities of the cable and deck systems upon the intensity of uniform distribution of live load to the cables was explored as early as 1877 by Professor W. Ritter (1877) of the Riga Polytechnical Institute, and A. Ritter (1862, 1879) indicates the principles involved. This elastic theory of suspension bridges implies small displacements and, therefore, high girder stiffness. Such high stiffness was regarded as essential to the avoidance of dangerous oscillation and became a feature of major suspension bridges generally, as the century came to an end. For these bridges the elastic theory provides a satisfactory basis of design. Fränkel acknowledges Müller-Breslau's contribution to the elastic theory (1881, 1886b); and his own treatment (1882), using the least work principle, is especially noteworthy. It is described in detail in Chapter 9.

Levy's version (1886) of the elastic theory is perhaps the most exhaustive. Apparently ignorant of the efforts of W. Ritter, Müller-Breslau and Fränkel, he begins with a critical examination of the then well-known Rankine theory. Then he proceeds by analysing the relationship of small, vertical deflexions of a suspension cable or chain to its overall length, having regard to small changes due to elastic strain and variation in temperature. After neglecting what he believes to be small quantities of the second and higher orders, he obtains the equation of compatibility of vertical deflexions of the chains with those of the deck or stiffening girders in bending, having regard to the configuration shown in Fig. 22 (his Fig. 1). That equation and the nature of the subsequent analysis is given

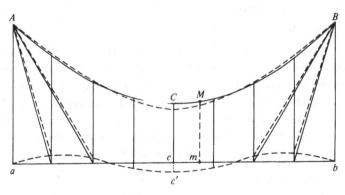

Fig. 22. From Levy (1886).

below: it is a matter for regret that Levy and Fränkel did not compare final results, for the latter seems to have achieved the same objective with relatively little effort. Thus, Levy writes:

$$\frac{1}{E}\int_{x_0}^{x_1} \frac{M}{I}(z-z_1)\,\mathrm{d}x = \left(\tau\delta + \frac{Q}{E_0 S_0}\right)l - (z_1 - z_0)\left(\frac{\mathrm{d}y}{\mathrm{d}x}\right) \qquad [(3.20)]$$

where:

E, E_0 are the moduli of elasticity of the deck and cables respectively;

x, x_1, x_0 are abscissae, the subscripts denoting the opposite ends of the relevant span distant l apart;

z, z_1, z_0 are ordinates of the cables with subscripts denoting terminal points;

y is vertical deflexion of cables and deck;

M is bending moment in the deck;

Q, S_0 are horizontal components of cable tension and cable cross-sectional area respectively;

δ, τ are coefficient of linear expansion and change of temperature respectively, for the cables.

Noting that $z_0 = z_1 = 0$ if the origin is chosen appropriately, Levy writes:

$$\int_0^l \frac{M}{I} z \, dx = \left(E\tau\delta + \frac{EQ}{E_0 S_0} \right) l \qquad\qquad [(3.21)]$$

and if strain changes due to changes in cable tension are negligible:

$$\int_0^l \frac{M}{I} z \, dx = E\tau\delta l \qquad\qquad [(3.22)]$$

In order to use equation (3.22) to determine the magnitude of the distributed reaction to live loads provided by the cables, Levy expresses the deck bending moment:

$$M = M_s - qm \qquad\qquad [(3.23)]$$

and substitutes in equation (3.22) to find q which represents the uniform intensity of the cable reaction. The other quantities represented are M_s, the bending moment in the deck as a simply supported beam, due to the live load; and the bending moment in the deck as a simply supported beam over span l, due to a uniformly distributed upward force (reaction) of unit intensity. Thus if I is uniform and there is no significant temperature variation, $\tau = 0$ and equation (3.22) may be written:

$$\int_0^l (M_s - qm) z \, dx = 0 \qquad\qquad [(3.24)]$$

whence:

$$q = \frac{\displaystyle\int_0^l M_s z \, dx}{\displaystyle\int_0^l m z \, dx} \qquad\qquad [(3.25)]$$

For a concentrated live load P distant α from the left and assuming the shape of the cables with dip f is parabolic, such that $z = 4fx(l-x)/l^2$, Levy gives:

$$q = \frac{5P}{l} \cdot \frac{\alpha}{l}\left(1-\frac{\alpha}{l}\right)\left(1+\frac{\alpha}{l}-\frac{\alpha^2}{l^2}\right) \qquad [(3.26)]$$

When $\alpha = l/2$ he notes that $q = 25P/16l$, which is substantially greater than the value of P/l used by Rankine for all positions of live load. Then at the other extreme when α is small, $q \ll P/l$. But the two theories agree for live load uniformly distributed over the whole span. Levy provided a theory which is *inter alia* appropriate for the interaction of inelastic cables with an elastic deck system on the basis that deflexions of the latter are small and, therefore, the cable shape is sensibly constant (that is, small deflexion of the cables without change in length).

A little-known but important contribution to the elastic theory of suspension bridges was made prior to Levy by Du Bois (1882), in the U.S.A., who makes no reference to any earlier attempts to solve the problem. After criticising Rankine's theory he continues 'we maintain in our present discussion that the curve of the cable does not remain parabolic but takes the curve of equilibrium due to the loading. We thus claim to obtain a more accurate, rational and scientific theory of the stiffening truss.' He said that the suspension system is appropriate for very long spans where the cable 'carries the dead weight in the most advantageous manner and, by reason of its own very considerable weight in such cases, resists in some degree the deforming action of partial loads'. He thus recognised gravity stiffness but neglected it subsequently. His approach is illustrated by Fig. 23 and is similar to that of Levy except for the terminal conditions of the truss. Those used by Levy are shown in Fig. 22 (effectively simply supported between the stiffening stays), while Du Bois believed that they should be encastré for increased stiffness.

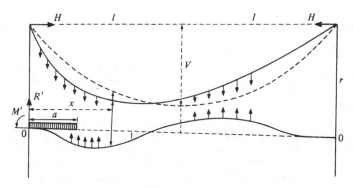

Fig. 23. From Du Bois (1882).

Ultimately, the important contribution from gravity stiffness which had been recognised in principle, implicitly to some extent since 1855, was taken into account by Melan (1888) and Godard (1894) and the deflexion theory of suspension bridges came into being. It appears that the true distribution of live load to the cable system is governed by the non-linear equation:

$$\frac{d^2q}{dx^2} = k^2q \tag{3.27}$$

where q is the (varying) intensity of that part of the total distributed load on the cable, due to the live load at a point x from the pier; $k^2 = (H+h)/EI$, where H is the horizontal component of the tension in the cables, due to the dead load; and h is the increment caused by the live load; while EI is the flexural rigidity of the truss. The solution gives results which agree with those of Levy when I is infinitely large or when deflexions are negligible and the number $l\sqrt{(H/EI)}$ is zero. Also, it seems that the elastic theory gives acceptable approximate solutions as long as that number does not exceed 5, where l is the span. Owing to the complexity of the theory, its development for practical utility belongs to the twentieth century, although Godard explored a variety of solutions and provided data and tables for design purposes.

Melan (1888) was also concerned with the theory of arches to allow for deflexion when the span is great and the arch is such that deformation, due to axial force and bending, cannot be neglected for purposes of design. Indeed, he considered the deflexion theories of the suspension bridge and the elastic arch as complementary. Useful descriptions of these theories are given by Am Ende (1898) and Balet (1908), the latter acknowledging the assistance of Melan. (Am Ende also describes the elastic theory using the least work method.)

Notes

Concerning the practical stability of arches, Moseley writes in his book (1843, p. 465):

So great is the *limiting angle of resistance* in respect to all the kinds of stone used in the construction of arches, that it would perhaps be *difficult* to construct an arch, the resultant pressure line upon any of the joints of which above the springing should lie without this angle, or which should yield by the slipping of any of its voussoirs.

The paper by Martin (1879) is remarkable as an example of an early British contribution (after Bell) to establish the elastic theory as a means of obtaining thrust-lines of arches. It is the more so because the author was a student of the Institution of Civil Engineers at the time and later, as a member of the staff of the journal *Engineering*, he helped to popularise Castigliano's least work principle (1895) and became an acknowledged authority on the theory of the steam turbine.

A. E. Young (1898) acknowledges that the theory of the elastic arch has been 'completely investigated', referring notably to Castigliano's celebrated book (1879) and the paper by Martin (1879). His objective is to examine Rankine's version of the theory (1861) with its complexity and minor inaccuracy.

It is useful to compare the equations of the elastic theory of suspension bridges derived by Levy with those of Fränkel (Chapter 9) using the strain energy method. The similarity is evident especially if the cables are inextensible.

Fidler (1878) in a paper read in 1874, before the Royal Scottish Society of Arts, described the difficulty of realising the economic advantages of suspension bridges for long spans because of their liability to dangerous oscillation and their consequent unsuitability for railways. He refers to American success, however, and suggests means of stiffening without undue increase in self-weight. Rankine's theory is mentioned and Jenkin's method (1873) of dealing with statical indeterminacy is quoted in relation to the analogy between suspension bridges and arches. But earlier P. W. Barlow advocated the suspension system for railway bridges, making adverse comparison between the economy of the Niagara Bridge and the Britannia tubular bridge (1858, 1860). (See also Chapter 11.)

4

Elementary theory of frameworks: graphical statics

Bar frameworks for bridges and roofs, with timber as the material of construction, are of ancient origin. Indeed, the illustrations of such structures in early nineteenth-century works on theory of structures, for example Navier's *Leçons* (1826), include them within the scope of carpentry, and pictorial illustrations clearly indicate timberwork. The widespread adoption of iron framework for roof trusses, arches and bridge girders, coincided with the construction of railways (a direct consequence of the new iron age) and an urgent need for numerous bridges and large buildings. The first major iron lattice girder bridges as an alternative to solid and plate girders (by virtue of the economy afforded by reduction in self-weight) appeared, it is believed, soon after 1840. Dempsey (1864) gives an interesting account of the history of cast and wrought iron construction, noting that the first iron vessels (boats) were made in 1820–1 by Manby of Tipton. He believed that the rolled 'I' section was introduced by Kennedy and Vernon of Liverpool in 1844, angle section being of earlier origin, and that Fairbairn made plate girders as long ago as 1832.

Dempsey credits Smart with the invention of the diagonal lattice girder, referred to as the 'patent iron bridge', in 1824 but seems to believe that it was first used for a major railway bridge after 1840, on the Dublin and Drogheda Railway. That bridge was described by Hemans (1844). The lattice girder did not at first receive the approval of some leading engineers, including Brunel and Robert Stephenson. It was advocated, notably by Doyne (1851), in a paper which drew adverse criticism from the learned Wild (assistant to Stephenson for the Britannia Bridge).

The invention (in various instances) in the U.S.A., of the rational bridge truss, by Howe, Jones, Linville, Whipple and others; along with Warren's invention of it in Britain (*c.* 1846), established the open framework for bridge construction with economy and safety. Bouch, with the help of Bow, used the so-called 'double Warren framework' for railway bridges in Britain from 1854 (Bow, 1855, 1873). Also, according to Heinzerling (c. 1873) and Straub (1952), the Belgian, Neville, used iron trusses with equilateral triangulation as early as 1845.

Theory of frameworks: Navier, 1826

The statics of frameworks was understood in time for the introduction of the iron truss for major structures. Thus, in Navier's *Leçons* the problems of equilibrium and deflexion of bar structures are illustrated

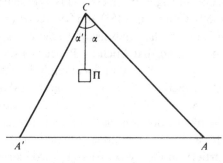

Fig. 24

in simple terms. Using Navier's method (and notation) for the plane two-bar structure shown in Fig. 24, first for the equilibrium of joint C at the apex:

$$\left.\begin{array}{l} p \cos \alpha + p' \cos \alpha' = \Pi \\ p \sin \alpha + p' \sin \alpha' = 0 \end{array}\right\} \tag{4.1}$$

where p and p' are the forces (assumed compressive) in the bars AC and $A'C$ respectively. Then the elastic small changes in length of the bars:

$$\left.\begin{array}{l} \text{for } AC: f \cos \alpha - h \sin \alpha \\ \text{for } AC': f \cos \alpha' - h \sin \alpha' \end{array}\right\} \tag{4.2}$$

are found by resolving the orthogonal components of deflexion h and f of C in the directions of the axes of the bars. By using the law of elasticity, relating the forces in the bars to their consequent changes in length, the deflexions h and f may be found, therefore, by these equations.

The necessary conditions for statical determinacy of a pin-jointed bar framework are implicit in Navier's analysis, but the underlying theory was considered explicitly by Möbius (1837), Maxwell (1864a), Levy (1874) and others and resulted in the well-known criteria of $2n-3$ bars and $3n-6$ respectively, bars for plane and space frameworks having n joints.

In Britain, Robison (Brewster, 1822) and, more especially, Whewell, were notable among contributors to the subject in the early years of the nineteenth century. (At this time Poncelet was professor of mechanics at the military academy of his native city of Metz.) Whewell, mathematician and sometime Master of Trinity College, Cambridge, included the application of elementary analytical statics to simple frameworks, in his book (1834) addressed to the needs of engineering. He discussed, incidentally, the problem of the king post truss and the means of overcoming statical indeterminacy when, as usual, the king post force is resisted by bending of the tie beam. Others engaged in engineering, especially Whipple (1847) in the U.S.A., Jourawski (1850), Culmann (1851) and Schwedler (1851) in Europe, contributed extensively to the analysis of frameworks, using algebra and arithmetic; and it is said that Michon, Poncelet's successor at Metz, lectured on truss analysis *c*. 1848. Graphical analysis was to be accepted as an advance on those methods. Drawing and graphical display was congenial to engineers, especially by virtue of ready observation of errors or defects of design. Winkler's opinions in this respect are recorded by Timoshenko (1953, p. 316) and Mohr mentions graphical methods for framework analysis in his article on theory of wood and iron structures (1860).

Graphical analysis: Poncelet, Rankine, Maxwell and Culmann

Such was the enthusiasm for graphical analysis during the latter half of the century that Chalmers (1881) suggested that an ideal course of engineering mechanics would begin with projective geometry, the 'modern geometry', as founded by Poncelet in 1813 (1822), to be followed by geometrical statics due to Möbius, Cousinery's 'calcul par le trait' and the funicular polygon of Varignon (1687), as interpreted and developed by Culmann (which Mohr applied to the elastic line with regard to deflexion of structures). Chalmers also emphasised the underlying geometric concept of engineering designs, remarking that structures are geometric forms whose forces, governed by the laws of statics, act along geometric lines. Of the engineer he said, accordingly, 'it is natural that he strove to follow a train of geometric thought'. Indeed, Chalmers believed that geometric methods possess a much higher value than analytical methods in expanding the intellectual powers, a belief which was evidently shared by Favaro (1879).

Chalmers acknowledged the priority of Rankine (1858) and Maxwell

(1864*a*) with regard to the graphical analysis of framework structures, but he believed that their methods were traceable to the teaching of Möbius. Moreover, he said that: 'These two developments require, however, to be supplemented by Culmann's method of obtaining the two reacting forces, in the case of ordinary frames, more especially when the impressed forces are not parallel. By themselves they remained comparatively unfruitful.' It is impossible to avoid the impression that Chalmers was prejudiced.

Thus, it was not long after the adoption of the metal framework (including trusses) for structures that powerful graphical methods of analysis were developed on the basis of the triangle and polygon of forces. Bow (1873) quotes an example of a force diagram for a truss (Fig. 25; his Fig. 342(i)) 'by the late Mr C. H. Wild, C.E., which was shown to me in the year 1854 but was of much earlier date'. Rankine (1870) claimed

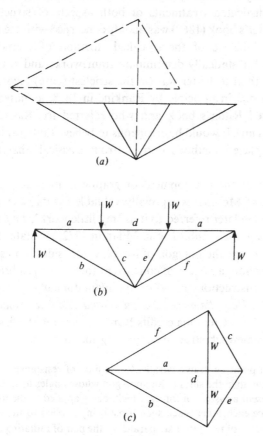

(*a*)

(*b*)

(*c*)

Fig. 25. From Bow (1873).

publication of force diagrams for frameworks, in lithographed notes for students at Glasgow University in 1856. Maxwell contributed his celebrated paper on 'reciprocal figures' regarding force diagrams in 1864 and Fleeming Jenkin (1869) gave a particularly lucid review of the subject and its origins. On the Continent, A. Ritter (1862), Cremona (1872) and Levy (1874), were also outstanding for their work in graphical statics of structures. But it is curious that Maxwell's important general paper on the subject (1870b), which included his theorem of least weight of frameworks (Charlton, 1963), was generally overlooked; a notable exception being Cotterill (1884).

It is interesting that innovation in graphical analysis of structures coincided with the beginning of intense activity in mathematical analysis which included analysis of statically-indeterminate bar frameworks, using the mathematical theory of elasticity. Indeed Levy's *La statique graphique* (1874) contains sophisticated treatments of both aspects of structural analysis. Also, A. Ritter's book (1862) was mainly concerned with the now well-known analytical device of the so-called 'method of moments' (method of sections) for statically-determinate frameworks and trusses. Attribution of the method to Ritter was (in the strictest sense) incorrect, for it had been published in principle by Rankine in 1858, as noted by Sankey who translated Ritter's book and who referred to 'Rankine's method of sections'! Thus, it would be erroneous to believe that graphical methods entirely displaced methods derived from classical analytical statics at any time.

Cremona's account of the development of graphical analysis (which includes the methods of Möbius) is especially valuable for its generality and historical review. He later referred to it as 'my little work', *Le figure reciproche nella statica grafica*, published in Milan in 1872. He states that the origins of the theory are the polygon of forces, whose sides represent, in magnitude and direction, a system of concurrent forces in equilibrium, and the geometrical constructions based on the plane funicular polygon due to Varignon (1725, *Nouvelle mécanique ou statique, dont le projet fut donné en 1687.* Paris). Then Cremona credits Rankine as being the first to apply the theory to framework structures, quoting the theorem

If lines radiating from a point be drawn parallel to the lines of resistance of the bars of a polygonal frame, then the sides of any polygon whose angles lie in these radiating lines will represent a system of forces, which, being applied to the joints of the frame, will balance each other; each such force being applied to the joint between the bars whose lines of resistance are parallel to the pair of radiating lines that enclose the side of the polygon of forces, representing the force in question.

Also, the lengths of the radiating lines will represent the stresses along the bars to whose lines of resistance they are respectively parallel.

It is noted that Rankine later (1864) published an analogous theorem for a system of polyhedral frames.

Cremona (1872) credits Maxwell with the geometrical theory of reciprocal diagrams. Maxwell, he said, defined them generally, and obtained them from the projections of two polyhedra which are reciprocal in respect of a paraboloid of revolution, in the manner of Poncelet's (or Monge's) theory of reciprocal polar figures. Cremona also notes that the practical application of the method of reciprocal figures was the subject of a memoir by Fleeming Jenkin, communicated in March 1869 to the Royal Society of Edinburgh.

Then Cremona acknowledges Culmann as 'the ingenious and esteemed creator of graphical statics' and notes that numerous questions on theoretical statics and other problems which relate to branches of practical science, are solved by his using a simple and uniform method which reduces itself in substance to the construction of two figures which he calls *Kräftepolygon* and *Seilpolygon* (polygon of forces and funicular polygon respectively). Cremona believed that Culmann did not consider these figures as reciprocal in Maxwell's sense but nevertheless they are substantially so; the geometrical constructions given by Culmann with regard to forces in frameworks almost invariably coincide with those derived by Maxwell's methods.

It is profitable here to quote Levy (1874):

It is, after all, it seems, a practitioner who had this first idea: M. Taylor, a simple mechanic of the English construction firm of J. B. Cochrane: but there are the works of Rankine (1857) and, above all, those of Clerk Maxwell (1864) on the theory of reciprocal figures which have given to Taylor's procedure the status of a precise and reliable method. As we write these lines we have from the family of M. Macquorn Rankine a letter telling of the death of the eminent professor of Glasgow University. That enables us to offer here our tribute and regret. His loss will be felt not only by men of science, but also by engineers and builders, for his researches have above all been of practical utility. His *Manual of applied mechanics* especially is the worthy successor of *Leçons sur mécanique industrielle* by Poncelet.

He who, with Rankine, Taylor and Maxwell, has contributed most to the development of graphical statics is Culmann by his teaching at l'Ecole Polytechnique de Zurich and by his major work, *Die Graphische Statik* of 1866.

Culmann did not exactly employ the theory of reciprocal figures of Maxwell: his figures are not always reciprocal and, for that reason, it may be feared that his methods were not used in so successful a fashion by his followers as by himself – by way of compensation, Culmann could treat certain problems to which reciprocal figures were not applicable.

Levy acknowledged Cremona's tract (1872) as presenting the theory of reciprocal figures, in a novel and elegant manner, as projections of reciprocal polyhedrals by Möbius' method for compounding forces in space. He suggested, moreover, that Maxwell's reciprocal figures are a direct consequence of a theorem due to Chasles. Then he names Poinsot, Coriolis, Binet, Dupin, Navier, Prony, Poncelet, Cauchy and Lamé as being equally distinguished in their use of geometrical methods in mechanics and that, contrary to popular belief, graphical statics is not of German origin. It was adopted there (toward 1870) but it seems that its initially slow development was probably due to the sophisticated approach used in presenting the new geometry (notably by Staudt) as a preliminary.

Levy gives particular credit to Bauschinger, professor at the Munich Polytechnikum, for his book published in 1871; and he also acknowledges Winkler, professor at Vienna, for his numerous papers on graphical statics. Emphasising the ease and rapidity of graphical methods in comparison with conventional methods, it is observed that while they might be lacking in decimal accuracy they are not subject to the gross errors of lengthy calculation 'where no aspect is readily visible' (Jenkin and Cremona make similar comments, as noted below).

But Mohr (1868) was sceptical about the value of graphical statics, having regard to the emergence of complicated structures and the prospects for the future (Chapter 2).

Maxwell, Jenkin and reciprocal figures

It is appropriate that Maxwell's contributions receive attention in some detail here. Thus, Jenkin's (1869) account includes the definition of reciprocal plane figures:

Two plane figures are reciprocal when they consist of an equal number of lines, so that corresponding lines in the two figures are parallel, and corresponding lines which converge to a point in one figure form a closed polygon in the other. If forces represented in magnitude by two lines of a figure be made to act between the extremities of corresponding lines of the reciprocal figure, then the points of the reciprocal figure will all be in equilibrium under the action of these forces.

Jenkin continues:

Few engineers would, however, suspect that the two paragraphs quoted put at their disposal a remarkably simple and accurate method of calculating the stresses in framework; and the author's attention was drawn to the method chiefly by the circumstance that it was independently discovered by a practical draughtsman, Mr Taylor, working in the office of the well-known contractor, Mr J. B. Cochrane. The object of the present paper is to explain how the principles above enunciated are to be applied to the calculation of the stresses in roofs and bridges of the unusual forms.

Cremona actually quoted these remarks (with the exception of the final sentence) and also from the penultimate paragraph of Jenkin's paper which continues:

When compared with algebraic methods, the simplicity and rapidity of execution of the graphic method is very striking; and algebraic methods applied to frames such as the Warren girders, in which there are numerous similar pieces, are found to result in frequent clerical errors, owing to the cumbrous notation which is necessary, and especially owing to the necessary distinction between odd and even diagonals.

The remainder of the paragraph remarks on the particular value of the graphic method when the loads are not symmetrical, and when they are inclined, as well as in cases such as the framed arch and suspension bridge. It also points out that the diagram, once drawn, acts as a sort of graphic formula for the strain on every part of a bridge or roof, and is a formula which can hardly be misapplied.

Jenkin notes that the construction of a reciprocal figure for any frame requires the exercise of a little discrimination and seeks to explain the method by examples: 'those frames only being considered which are so braced as to be stiff, but have not more members than is sufficient for this purpose'. His first example is that of a triangle loaded in the middle, and

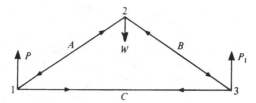

Fig. 26. From Jenkin (1869).

supported at the two ends, as shown in Fig. 26, and he describes the construction of the reciprocal diagram of forces shown in Fig. 27.

Rankine (1872) provides a particularly concise and lucid account of Maxwell's reciprocal figures which (he says) were devised in 1857, with reference to some roof trusses, as shown in Fig. 28, and mentions his own contribution (1858).

Examples treated by Jenkin are shown in Figs. 29 and 30. That shown in Fig. 30 is especially interesting, being essentially a statically-indeterminate framework, about which, he says, more members are used than suffice to

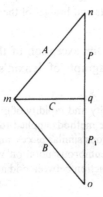

Fig. 27. From Jenkin (1869).

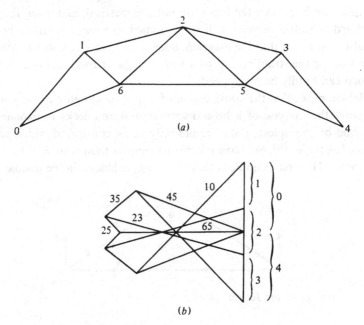

Fig. 28. From Rankine (1872).

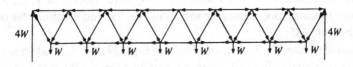

Fig. 29. From Jenkin (1869).

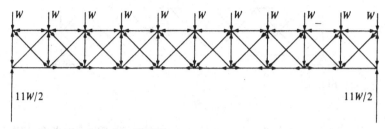

Fig. 30. From Jenkin (1869).

render the frame stiff, the stresses are indeterminate and the frame may be self-strained. He notes that:

If, however, the verticals be alone suited to resist the còmpression, the diagonals being fit to sustain tension only, the stresses become determinate, half the diagonals being with any given loading wholly inoperative. The reciprocal figure can be used to discover which are the active members, as they may be called, and what are the strains upon them.

On this basis Jenkin gives the reciprocal figure for a uniformly distributed load on the framework, together with the component polygons for that figure (he disregards the inactive diagonal members by omitting his customary numerals of designation).

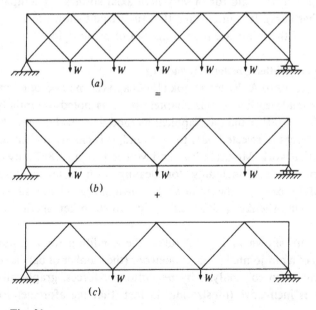

Fig. 31

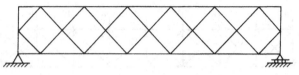

Fig. 32

He observes that the lines representing the external forces acting on a rigid frame in equilibrium must, in the reciprocal figure, form a closed polygon, and when these lines are parallel, as for loads only, the polygon becomes infinitely thin, and is represented by a single straight line, subdivided into parts proportional to the forces.

It is appropriate to take note here of similar treatment of another inherently statically-indeterminate framework (favoured particularly by Bouch) described clearly by Weyrauch (1888) and by Salmon (1938). The framework and method (which is approximate and depends upon symmetry and superposition (and is believed to be due to Jourawski, *c*. 1850), is illustrated in Fig. 31. The structure became commonly known as the 'double Warren girder'. For complete symmetry in respect of the proportions of bars and symmetrical loading, the approximate treatment illustrated is acceptable and is indeed accurate if the vertical bars at each end of the girder are rigid (or of very high axial stiffness in comparison with the other bars). It is interesting that the variant shown in Fig. 32 is a mechanism if the joints are pinned instead of being rigid.

Mohr, Müller-Breslau, Henneberg

Reference to A. Ritter's book (1862) and the method of moments (sections) is made earlier in this chapter. It is, as noted, essentially an analytical device rather than a graphical method and, in the sense that it enables the forces in selected bars to be found, it is closely related to the device of virtual work. Mohr (1874*a*), and Müller-Breslau (1887*a*) used the latter to great effect, including for dealing with instances (usually hypothetical, for example the framework shown in Fig. 33, and similar to that which Timoshenko (1953) cited) in which other methods are unsuitable.

The structure shown in Fig. 33 fulfils the condition for the statical determinacy of a pin-jointed plane framework (the number of bars is equal to $2n - 3$) but, even so, analysis by resolution of forces, graphically or analytically, is ineffective (illustrating, in fact, that the aforementioned

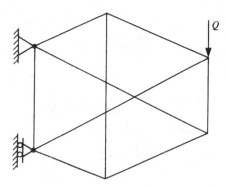

Fig. 33. After Henneberg (1886).

condition for statical determinacy is necessary but insufficient by itself). If any bar is removed, however, the system is transformed into a collection of four-bar chain mechanisms since there are, *ex hypothesis*, no joints where bars cross over one another (those crossings being assumed perfectly free). The virtual work method consists, therefore, in removing a bar and substituting the forces which it would supply at its terminal joints to maintain the resulting (loaded) mechanism in equilibrium. Considering a small compatible displacement of the system, involving the actual loads and the forces substituted for the missing bar, provides the work equation whereby the latter may be found. The remaining bar forces may then be found by elementary statics. The particular contributions of Müller-Breslau appear to be related to the treatment of the kinematics of the system, having selected a bar for removal and determining its force by virtual work.

Henneberg's method (1886) of dealing with structures of the kind shown in Fig. 33 is especially ingenious. It depends upon the device of varying the arrangement of bars within the overall pattern defined by the positions of the joints; for example by temporarily removing one bar and introducing another elsewhere to prevent the system from becoming a mechanism (Fig. 34(*a*)). Having done that, the modified system is analysed for the specified loading and the forces in every bar are determined. The loading is then removed while unit forces are applied across the gap (Fig. 34(*b*)) caused by the removed bar, and the resulting forces in the bars are calculated. A multiplying factor may then be calculated for the latter analysis in order to make the force in the substitute bar equal and opposite to the force induced in it by the original or specified loading of the structure. Superposition of the two analyses gives the forces in the bars

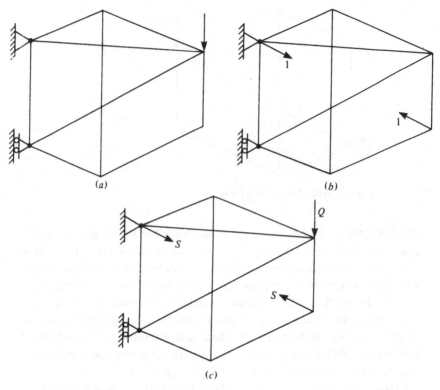

Fig. 34. After Henneberg (1886).

of the specified or original structure and loading as shown in Fig. 33. According to Timoshenko (1953), Saviotti (1875, 1888) and Schur (1895) provided other general methods of solving frameworks of this kind.

Space frameworks

Möbius (1837) is believed to have been the earliest contributor to the rigorous analysis of space frameworks (including general conditions of rigidity, for example that the determinant of the equations of equilibrium is non-zero), and his work seemed to be overlooked by engineers. Also, the mathematician Clebsch (1862) includes a space framework in his analysis of elastic systems (described in detail in Chapter 5). But A. Föppl seems to have been distinguished for being among the first engineers to deal with this aspect in general terms, while expressing surprise that the general theory of space frameworks should be unknown (to engineers) in 1891. Schwedler's cupola, a particular space framework, was, however, known in 1866 and is among the structures considered by Föppl in his book (1892) which was concerned mainly with trusses and the design of metal bridges.

The device of tension coefficients for space frameworks, so apparent and yet elusive, did not become well known in Britain until the twentieth century, though it is believed to have been used by Müller-Breslau (1892) in the first instance. The concept of tension coefficients may be deduced from Clebsch's equations (1862) for the equilibrium of a joint of a space framework. Thus, with external orthogonal forces or loads X_i, Y_i and Z_i applied to the ith joint of the framework:

$$\left. \begin{aligned} X_i + \sum_j T_{ij} \frac{(x_j - x_i)}{r_{ij}} &= 0 \\ Y_i + \sum_j T_{ij} \frac{(y_j - y_i)}{r_{ij}} &= 0 \\ Z_i + \sum_j T_{ij} \frac{(z_j - z_i)}{r_{ij}} &= 0 \end{aligned} \right\} \qquad [(4.3)]$$

where T_{ij} is the force (tension) in any bar connecting the particular joint to any other joint denoted by j; r_{ij} is the length of that bar and x_i, x_j; y_i, y_j; z_i, z_j are the coordinates of the joints i and j. Defining T_{ij}/r_{ij} as the tension coefficient t_{ij}, the equations may be rewritten:

$$\left. \begin{aligned} X_i + \sum_j t_{ij}(x_j - x_i) &= 0 \\ Y_i + \sum_j t_{ij}(y_j - y_i) &= 0 \\ Z_i + \sum_j t_{ij}(z_j - z_i) &= 0 \end{aligned} \right\} \qquad (4.4)$$

and having found the values of the tension coefficients for a statically-determinate framework by using these equations, the force in any particular bar may be found simply by multiplying the relevant coefficient by the length of the bar. It is, therefore, necessary to calculate only the lengths of those bars whose forces are required and that may well result in substantial saving of labour for a complicated space structure.

Deflexion of frameworks: Williot, Mohr

The theory of the calculation of the deflexion of joints of frameworks was clearly embodied in Navier's elegant method of analysis (1826), as described briefly above. But it was not until the development of major metal structures (which called for a solution to the problem of statical indeterminacy in frameworks) that the subject was attacked by Maxwell, Mohr and others (after 1864). The lead which had been provided by Navier had been disregarded (except perhaps by Levy) and the elastic energy theory, as well as the principle of virtual work (virtual velocities), had been used instead to develop special methods. They had the advantage of enabling the deflexions of selected joints to be found individually and are described in Chapters 5 and 8, the latter being devoted to energy concepts and methods in theory of structures. Mohr's method of finding the deflexion of joints of chords of bridge trusses, by analogy with bending moments in a simply supported beam, is also noteworthy and is described in Chapter 10. But the method which, perhaps, achieved the greatest popularity for estimating the deflexions of frameworks, is the well-known graphical method due to the French engineer Williot (1877*b*), which featured in standard texts on theory of structures (for example Salmon, 1938). By its very nature it is complementary to graphical analysis of forces in frameworks. Indeed, it is the graphical interpretation of the equations of Navier relating deflexions of joints with the small elastic changes in the length of the bars of a framework, on the basis of constant geometry. The method is explained in Williot's words by Krohn (1884), using Figs. 35 and 36, in an article devoted to an elegant application of

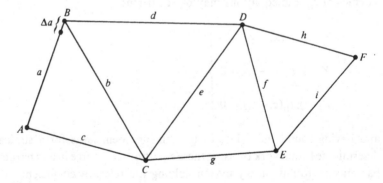

Fig. 35. From Krohn (1884).

the law of reciprocal displacements for analysing a two-pin framed arch. Assuming that displacements are referred to the line *AB* for the framework shown in Fig. 35, with *A* as the fixed point, the vector diagram of displacements of joints compatible with small changes in length $\Delta a, \Delta b, \Delta c, \ldots$, of the bars is constructed as shown in Fig. 36. The solid lines represent those changes in length and the broken lines represent the

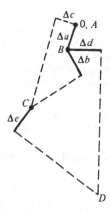

Fig. 36. From Krohn (1884).

effect of consequential small rotations of bars. (The diagram is incomplete: joints *E* and *F* are not represented.)

Mohr (1887) supplemented Williot's method by providing a graphical device for referring displacements to a fixed datum rather than one determined at the conclusion of the graphical analysis. Thus, with regard to vertical deflexions of frameworks supported simply at two opposite ends, the Williot diagram used a point datum at one end of the structure and it is found that vertical deflexion of the opposite end is implied by the diagram. In fact, the vertical deflexion of the intermediate points of the structure are determined by a line between the fixed support point and the position of the other support implied by the diagram. Mohr provided a graphical means of referring all such deflexions to a horizontal datum and when this is applied the complete deflexion diagram is known as the Williot–Mohr diagram.

An important aspect of the design of statically-determinate frameworks concerns the matter of arrangement of bars (that is, the general form of a framework) for a specified function, so that the least amount of material is necessary. This is particularly relevant for major bridge structures, both in the interests of elementary economy and the very significant factor of self-weight which is indeed dominant in many bridge projects. Levy included a very useful study of this subject in his celebrated book (1874) and showed (Chapter 6) that for a specified loading and parallel chords, the truss consisting of equilateral triangles (Warren girder) is potentially the best.

Notes

Jourawski used analytical statics for the design of railway bridge trusses as long ago as 1844, according to Timoshenko (1953), with particular reference to the Howe truss. It is believed that he originated the use of superposition of constituent statically-determinate systems, by symmetry, for the analysis of such trusses with supernumary bars, a technique which became established and was used to advantage, notably by Weyrauch (1887).

Weyrauch made a thorough study and appraisal of statically-determinate systems for bridges (1887, 1888) at a time when others were preoccupied with the theory of statically-indeterminate structures.

The use of kinematics to facilitate framework analysis is explained in detail by Land (1888) in an article on the subject.

A. Jay Du Bois (1875–7) seems to share with Chalmers (1881) the distinction of providing the English speaking world with a comprehensive account of the German development of the major discipline of graphical statics, embracing, *inter alia*, the 'elastic line' and continuous beam theory. Its limited development in Britain and France for statics, in the strictest sense only, is noted. There are comprehensive bibliographies, with annotations, and Weyrauch's (1874*a*) history is reproduced, noting the controversy over the value of von Staud's 'modern geometry'.

5

Theory of statically-indeterminate frameworks: the reciprocal theorem

The precise analysis of statically-indeterminate systems of bars, including trusses and pin-jointed frameworks generally, seems to be due to the famous French engineer, Navier. It was included in his lectures at l'Ecole des Ponts et Chaussées, which appeared in the form of his celebrated *Leçons* in 1826. According to Saint-Venant (Navier, 1864, p. 108) the method was part of the course as early as 1819. It was elaborated (1862) by the mathematician Clebsch in Germany; while, in Britain, Maxwell (1864*b*) who, it seems, was unaware of Navier's elegant and general method, published an original method of solving the problem. Levy, who was apparently aware of Navier's work, published a novel method in 1874 (Chapter 6). But it was not really until the German engineer, Mohr, published his analysis in the same year that the subject began to be appreciated by engineers (on the Continent at first and much later in Britain).

This chapter is concerned with those original contributions, in principle only: various sophistications and devices to increase their utility in engineering are considered in Chapters 8 and 10.

Navier, 1826

Navier's contribution to the analysis of statically-indeterminate pin-jointed systems is to be found essentially in the two articles of his *Leçons* (1826, art. 632, p. 296; 1833, art. 533, p. 345). There he states: 'When a load is supported by more than two inclined bars in the same vertical plane or by more than three inclined bars not in the same plane, the conditions of equilibrium leave undetermined, between certain limits, the forces imposed in the direction of each of the bars.' He continues with a discussion of the determination of the limits between which the member

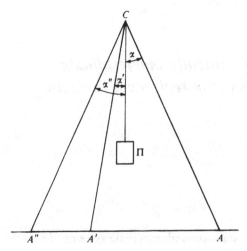

Fig. 37. From Navier (1826).

forces would lie, on the basis that the bars are rigid. Navier suggests that this be done by considering two bars to be effective at one time in the plane system and three at one time in the three-dimensional system. He then goes on to say that, in fact, bars are capable of deformation in a way dictated by the elasticity of their material and that, if account is taken of such deformation, the distribution of load between the bars is no longer indeterminate. Then, with reference to Fig. 37 (his Fig. 112), he says:

To give an example, suppose the weight Π is supported by three inclined bars AC, $A'C$, $A''C$ in the same vertical plane, and that we call α, α' α'' the angles formed by the directions of the three bars with the vertical cord $C\Pi$; p, p', p'' the forces exerted, due to the action of the weight Π, in the directions of the bars; F, F', F'' the elastic forces of the three bars; a the height of the point C above the horizontal line AA''; h, f the amounts by which the point C is displaced, horizontally and vertically (down), by the effect of the simultaneous compression of the three bars.

(Denoting the elastic force of the bar AC by F, means that it would need a force of F to elongate or shorten this bar by an amount equal to its actual length, and similarly for the other bars.) Navier then writes the equations of equilibrium:

$$\left. \begin{array}{l} p \cos \alpha + p' \cos \alpha' + p'' \cos \alpha'' = \Pi \\ p \sin \alpha + p' \sin \alpha' + p'' \sin \alpha'' = 0 \end{array} \right\} \qquad [(5.1)]$$

and the changes in length of AC, $A'C$ and $A''C$ as $f \cos \alpha - h \sin \alpha$; $f \cos \alpha' - h \sin \alpha'$; $f \cos \alpha'' - h \sin \alpha''$; assuming them to be very small.

Noting that the lengths of the bars are:

$$\left. \begin{array}{l} a/\cos\alpha \\ a/\cos\alpha' \\ a/\cos\alpha'' \end{array} \right\} \qquad [(5.2)]$$

respectively, he writes the fractions of their lengths represented by their compression as:

$$\left. \begin{array}{l} (f\cos^2\alpha - h\sin\alpha\cos\alpha)/a \\ (f\cos^2\alpha' - h\sin\alpha'\cos\alpha')/a \\ (f\cos^2\alpha' - h\sin\alpha''\cos\alpha'')/a \end{array} \right\} \qquad [(5.3)]$$

Finally, he writes the expressions:

$$\left. \begin{array}{l} p = F(f\cos^2\alpha - h\sin\alpha\cos\alpha)/a \\ p' = F'(f\cos^2\alpha' - h\sin\alpha'\cos\alpha')/a \\ p'' = F''(f\cos^2\alpha'' - h\sin\alpha''\cos\alpha'')/a \end{array} \right\} \qquad [(5.4)]$$

and says: 'which, together with the two equations above, will give the values of the displacements h and f, and the forces p, p' and p'' '.

It is noteworthy that, assuming the forces p, p' and p'' are compressive, the first of Navier's equilibrium equations is correct but there is an error of sign in the second. Thus the term relating to p should be negative. This does not, however, detract from the elegance and clarity of Navier's exposition as a whole.

Clebsch, 1862

Clebsch, adopting Navier's approach, treats frames (systems of bars) with linear elasticity, in the final chapter of his book (1862) on the theory of elasticity of solid bodies; in fact, for this reason, it seems to end on a diminutive note after the welter of complicated mathematics of the earlier chapters – and without a single diagram! Thus article 90 of Clebsch's book is concerned with 'Systems of bars without bending'.

In this article he derives the following general equations for the equilibrium of the ith joint of any freely jointed space framework:

$$\left. \begin{array}{l} X_i + \sum_j E_{ij}\sigma_{ij}\rho_{ij}(x_j - x_i)/r_{ij}^2 = 0 \\ Y_i + \sum_j E_{ij}\sigma_{ij}\rho_{ij}(y_j - y_i)/r_{ij}^2 = 0 \\ Z_i + \sum_j E_{ij}\sigma_{ij}\rho_{ij}(z_j - z_i)/r_{ij}^2 = 0 \end{array} \right\} \qquad [(5.5)]$$

where X_i, Y_i, Z_i are the loads applied to the joint; E_{ij} is the modulus of elasticity of the bar connecting any joint j to the particular joint i; r_{ij} is the unstrained length of the bar; σ_{ij} is the cross-sectional area of the bar; x, y, z, with appropriate subscripts, are the cartesian coordinates of joints of the framework; and ρ_{ij} is the small change in length of the bar (elastic deformation), which may be expressed in terms of the elastic displacements u_i, v_i, w_i; u_j, v_j, w_j of the joints i and j respectively, as:

$$\rho_{ij} = [(x_i - x_j)(u_i - u_j) + (y_i - y_j)(v_i - v_j) + (z_i - z_j)(w_i - w_j)]/r_{ij} \qquad [(5.6)]$$

Clebsch notes that if there are n joints in the framework, $3n$ equations are obtained, which are sufficient for the determination of the $3n$ unknowns u, v and w (from which the forces in the bars can be found). Clebsch goes on to treat the simple example of a force acting upon the common joint of a number of bars, the other ends of which are fixed. He takes the origin of the cartesian coordinates at the common joint and for the jth member writes for its small change in length:

$$\rho_j = -(ux_j + vy_j + wz_j)/r_j \qquad [(5.7)]$$

He proceeds to express the three equations of equilibrium of the system as follows:

$$\left.\begin{aligned} X &= a_{xx}u + a_{xy}v + a_{xz}w \\ Y &= a_{yx}u + a_{yy}v + a_{yz}w \\ Z &= a_{zx}u + a_{zy}v + a_{zz}w \end{aligned}\right\} \qquad [(5.8)]$$

and defines the coefficients as:

$$\left.\begin{aligned} a_{xx} &= E_j \sigma_j x_j^2/r_j^3 \\ a_{xy} &= a_{yx} = E_j \sigma_j x_j y_j/r_j^3 \\ a_{yy} &= E_j \sigma_j y_j^2/r_j^3 \end{aligned}\right\} \qquad [(5.9)]$$

The complete solution of the problem is given by the equations:

$$\left.\begin{aligned} u &= (\Delta_{xx}X + \Delta_{xy}Y + \Delta_{xz}Z)/\Delta \\ v &= (\Delta_{yx}X + \Delta_{yy}Y + \Delta_{yz}Z)/\Delta \\ w &= (\Delta_{zx}X + \Delta_{zy}Y + \Delta_{zz}Z)/\Delta \end{aligned}\right\} \qquad [(5.10)]$$

It is interesting to note the generality of Clebsch's approach for linear systems, including his systematisation of the analysis by introducing (stiffness) coefficients of elasticity of a structure and their determinants Δ (he noted the reciprocal property some two years before the publication of Maxwell's statement of the reciprocal theorem). In fact, he (and Navier) adopted that very approach, the 'equilibrium approach' (the inverse of that used later by Maxwell and Mohr) which has been found so convenient for the complete programming of structural analysis.

Maxwell and Jenkin, 1864–1870

In essence Navier proposed (as did Clebsch) a general method for the resolution of statically-indeterminate structures, based upon the necessary conditions of equilibrium of forces and compatibility of deformations, and the law of elasticity. However, it seems that he did not pursue its general possibilities; and though it was to be rediscovered and widely used in the twentieth century (by virtue of the nature of much modern construction technology), it seems to have made little or no impression during the nineteenth century, for it was apparently unknown to Jenkin and Maxwell in Britain, and to Mohr in Germany. Among its features is one whereby the number of equations to be solved finally depends upon the number of independent components of displacements of joints ('degrees of freedom') that are necessary to specify the deflexion of a structure caused by specified loads. Whereas in those modern structures with rigid connections there are usually many more statically-supernumerary (redundant) elements (including connections as well as bars) than degrees of freedom of deflexion, the converse is the rule for triangulated frame construction which predominated in the times of Maxwell and Jenkin. Thus, in any event, Navier's method tended to be unduly laborious for the latter kind of construction and, among engineers, Mohr became the acknowledged leader in the field, with his method which was similar in principle to Maxwell's, but was devised independently some ten years later.

In 1861, in his capacity as a consulting engineer, Jenkin had occasion to study the structural problem of bridge design. Thus, in 1869, he wrote concerning the problem of an arch rib or braced chain connected to abutments by pin-joints:

The direction of the thrust at the abutments is indeterminate until we have taken into account the stiffness of the rib and yield of the abutments...No stress in nature is, however, really indeterminate, and a method of calculation introduced by Professor Clerk Maxwell enables us to solve this problem, and the solution has led to remarkable conclusions.

The author in 1861 perceived that the true form of a stiff rib or chain would be that in which two members should be braced together as in the girder; but whereas in the girder one member was in compression and the other in tension, in the braced arch both members might be compressed, and in the suspension bridge both extended. The only effect of varying the distribution of load on such structures would be, that the thrust would at one time be such as to throw the chief strain on the upper, and at another time on the lower member. In this way a stiff frame could be produced which was essentially an arch or suspension bridge, differing from the girder, which essentially contains one member which shall be extended and one which shall be compressed.

He found, however, that he was unable to calculate, except on unproved assumptions, the true distribution of strains on the framework, being unable to determine the resultant thrust; but having drawn Professor Maxwell's attention to the problem, he [Maxwell] discovered and published, in the *Philosophical Magazine* in May 1864 a method by which the resultant and all stresses in framed structures could be positively determined.

Maxwell's paper (1864*b*) is remarkable for the physical insight displayed, its brevity (without diagrams), breadth and originality. It begins:

The theory of the equilibrium and deflections of frameworks subjected to the action of forces is sometimes considered as more complicated than it really is, especially in cases in which the framework is not simply stiff, but is strengthened (or weakened as it may be) by additional connecting pieces.

Maxwell appears to assume it will be understood that he is addressing himself to the problem of frameworks in which the bars ('pieces', in Maxwell's terminology) are pin-jointed together to form connected triangular shapes that are capable of sustaining simple axial force only. Having said 'I have therefore stated a general method of solving all such questions in the least complicated manner', he mentions that his method is derived from the well-known principle of conservation of energy referred to by Lamé (1852) as 'Clapeyron's Theorem' (Chapter 7). Then Maxwell discusses, in very brief and general terms, the necessity of 'equations of forces', 'equations of extensions' and 'equations of elasticity' for the solution of the general problem in three dimensions, and for the geometrical definition of a frame. Subsequent revelation of the details of his approach to the problem demonstrates an ingenious physical concept of matching the deformations of the simply-stiff (statically-determinate) framework to those of its supernumerary (redundant) bars which result from the application of external forces. That process, in which equilibrium of forces is satisfied at each stage, finally provides equations of deformation. These are as numerous as those bars which are additional to the requirements of statics.

In order to determine deformations of the simply-stiff framework, Maxwell (1864*b*) applies the law of conservation of energy (Clapeyron's theorem) to the work done against elasticity, and assumes linearly elastic behaviour. At first he states the following theorem which apparently refers to a hypothetical framework:

If p be the tension of the piece A due to a tension-unity between the points B and C, then an extension-unity taking place in A will bring B and C nearer by a distance p.

For let X be the tension and x the extension of A, Y the tension and y the extension of the line BC; then supposing all other pieces inextensible, no work will be done except in stretching A, or

$$\tfrac{1}{2}Xx + \tfrac{1}{2}Yy = 0 \qquad\qquad [(5.11)]$$

But $X = pY$, therefore $y = -px$, which was to be proved.

An interesting reciprocal property of linearity thus emerges. Maxwell's reasoning, however, clearly refers to a simply-stiff linearly elastic framework (Fig. 38(a)), otherwise it would break down with the assumption that all pieces are inextensible except A.

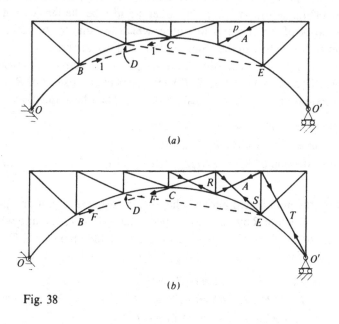

(a)

(b)

Fig. 38

Then he proceeds to consider the problem of the effect of a tension applied between the points B and C of a simply-stiff framework (such as that shown in Figure 38(a)) in respect of the extension of the line between two other points D and E if all bars except A are inextensible. Having found the tension in each bar due to unit tension between B and C and designating that in A by p, the process is repeated for unit tension between D and E, designating that in A by q. Then if the tension between B and C is F, the tension in A is Fp, and if the flexibility of that bar is e, its extension is eFp and by the previous theorem the corresponding extension of the line DE is $-Fepq$. Maxwell notes that if all bars of the framework are extensible then the result is $-F\Sigma(epq)$, while for the special case of the line BC the extension of that line is $-F\Sigma(ep^2)$. These summations would nowadays be identified as the flexibility coefficients $a_{(BC)(DE)}$ and $a_{(BC)(BC)}$. Subsequently he remarks 'that p and q always enter into the equations in the same way, so that we may establish the following general theorem: the extension in BC, due to unity of tension along DE, is always equal to the tension (extension) in DE due to unity of tension in BC'. This is, in fact, the statement of the reciprocal theorem for linear systems, which bears Maxwell's name.

The second problem which Maxwell considers relates to the extension between D and E due to a tension F between B and C when the framework is not simply stiff but has additional bars $R, S, T, \ldots$, whose elasticities are known (Fig. 38(b)). He proceeds to let p and q be the tensions in the (typical) bar A of the simply-stiff frame, which are due to unit tensions in BC and DE (as before in the absence of the additional bars $R, S, T, \ldots$,) and to let $r, s, t, \ldots$, be the tensions in A due to the respective unit tensions in $R, S, T, \ldots$, and also to let R, S, T, be the respective tensions of R, S, T, and their extensibilities (flexibilities) be ρ, σ, τ. Then he states that:

$$\left.\begin{array}{l} \text{the tension of } A \text{ is } Fp + Rr + Ss + Tt + \ldots \\[4pt] \text{the extension of } A \text{ is } e(Fp + Rr + Ss + Tt + \ldots) \\[4pt] \text{the extension of } R \text{ is } -F\Sigma epr - R\Sigma er^2 - S\Sigma ers - T\Sigma ert + \ldots = R\rho \end{array}\right\}$$

$$[(5.12)]$$

(by the first theorem stated above, whereby extension–unity of any piece A will shorten the line R by r in the absence of bars $R, S, T, \ldots$, and where the summations include all bars such as A but exclude bars $R, S, T, \ldots$, both of which features Maxwell did not explain); similarly the extension

$$\left.\begin{array}{l} \text{of } S \text{ is } -F\Sigma eps - R\Sigma ers - S\Sigma es^2 - T\Sigma est + \ldots = S\sigma \\[4pt] \text{of } T \text{ is } -F\Sigma ept - R\Sigma ert - S\Sigma est - T\Sigma et^2 + \ldots = T\tau \\[4pt] \text{of } DE \text{ is } -F\Sigma epq - R\Sigma eqr - S\Sigma eqs - T\Sigma eqt + \ldots = x \end{array}\right\}$$

$$[(5.13)]$$

noting that there are sufficient equations, first to determine the unknown quantities $R, S, T, \ldots$, and then to find the required extension between D and E.

Here the flexibility coefficients of linear elasticity may be identified and their reciprocal property noted, that is, $\Sigma er^2 = a_{rr}$; $\Sigma ers = a_{rs} = a_{sr}$; $\ldots$, though Maxwell did not develop his definitions and notation to that extent.

Finally, Maxwell proposes:

In structures acted on by weights in which we wish to determine the deflexion of any point, we may regard the points of support as the extremities of pieces connecting the structure with the centre of the earth; and if the supports are capable of resisting a horizontal thrust, we must suppose them connected by a piece of equivalent elasticity. The deflexion is then the shortening of a piece extending from the given point to the centre of the earth.

Though the kind of approach adopted by Maxwell had, it seems, been used by others for particular problems, such as beams on several rigid supports and curved bars restrained at each end (Navier, 1826 and Bresse, 1854), the more difficult and general problem posed by the statically-indeterminate framework had not hitherto been treated in that particular way.

Due perhaps to its brevity, absence of pictorial illustration and a number of clerical errors, Maxwell's paper did not, it seems, attract much attention at first. The publication in 1873, however, of Jenkin's contribution of 1869 to the subject, related as it was to specific practical problems, probably stimulated the interest of some engineers, for at that time Jenkin was the first incumbent of the Regius Chair of Engineering at the University of Edinburgh, having been appointed in 1868. (Maxwell occupied the Chair of Natural Philosophy at King's College, London in 1864 when his paper appeared.)

Jenkin, 1869

The importance of Jenkin's paper resides mainly in his introduction to the subject of structural analysis of a way of using the concept of conservation of energy to calculate deflexions, whether due to elasticity or otherwise, namely, the ancient principle of virtual work ('virtual velocities', as it was then called, as noted in Chapter 7). Indeed, Jenkin said with reference to Maxwell's method: 'The following is an abstract of the method and of the reasoning by which it is established, put into a form in which it will be more readily understood and applied by practical men.' The particular application which Jenkin considered was a symmetrical arch constructed as a triangulated frame which is connected by pin-joints to rigid, in-line abutments, whereby the horizontal component of the thrust of an abutment, due to loading of the structure, was statically-indeterminate. He noted that the single equation of compatibility of deformation, which enabled that thrust to be found, specified zero horizontal deflexion of the frame in the line of the abutments. In the paper Jenkin uses the principal of virtual velocities and so first derives an expression for that horizontal deflexion which is due to a small change in length, or strain, in a single bar. Thus, denoting the force in that bar, due to unit horizontal force acting on the frame at an abutment O, by q (Fig. 39(a)); a small change in length

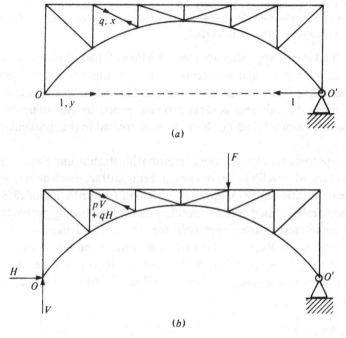

Fig. 39. After Jenkin (1873).

of the bar by x; and the corresponding horizontal deflexion of the frame in the line of the abutments by y, he writes:

$$1y = qx \qquad [(5.14)]$$

He then specifies that unit vertical force acting upward at an abutment, as the result of a certain load (Fig. 39(*b*)) on the frame, corresponds with a contribution p to the force in the chosen single bar so that, for horizontal and vertical forces H and V acting on the frame in equilibrium at an abutment, the force in the bar is $pV + qH$. Thus if e is the extension or compression of the bar due to unit force in it, then the value of x due to V and H is

$$x = e(pV + qH) \qquad [(5.15)]$$

and, by the principle of virtual velocities, the corresponding value of y is

$$y = epqV + eq^2H \qquad [(5.16)]$$

If now all bars of the frame are capable of deformation, the total horizontal deflexion in the line of the abutments associated with V and H is

$$\Sigma y = V\Sigma epq + H\Sigma eq^2 \qquad [(5.17)]$$

and if the abutments are rigid $\Sigma y = 0$, whence:

$$H = V\Sigma epq/\Sigma eq^2 \text{ numerically} \qquad [(5.18)]$$

As V is determined by statics, the problem of finding H is solved, though a specific sign convention is necessary for a numerical calculation.

Jenkin proceeds to consider both the effect of elastic yield of abutments and of thermal expansion, and finally gives details of numerical calculations for a specific frame, set out systematically in tabular form, as part of the study of weight saving with which the paper is mainly concerned.

Unfortunately, there is no evidence to show that Jenkin's efforts to simplify Maxwell's method by introducing the principle of virtual velocities appealed to 'practical men'. It was, however, with the aid of that device that Mohr developed his method of framework analysis, which was identical in principle to Maxwell's. In Britain, the enormous value of the principle of virtual work in theory of structures has been appreciated widely but only comparatively recently. As may be judged from Jenkin's work, it is independent of appeal to any law of elasticity, being concerned solely with systems of forces in equilibrium and with systems of displacements compatible with the geometry of the force systems.

Mohr, 1874

In 1874 Mohr addressed himself to the problem of the linearly elastic structure shown in Fig. 40 (Figs. 1 and 2 of his article, 1874*a*), which is practically identical to that which Jenkin chose to solve, using Maxwell's

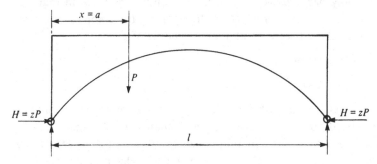

Fig. 40. From Mohr (1874*a*).

method simplified by introducing the principle of virtual work. Mohr dealt with the statics of the problem by A. Ritter's methods taken (as he acknowledged) from the book *Elementare Theorie der Dach- und Brücken-Konstructionen* (1862), but first he noted that the problem could be

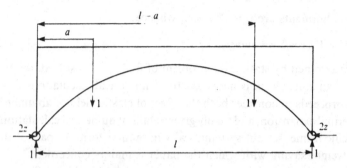

Fig. 41. From Mohr (1874*a*).

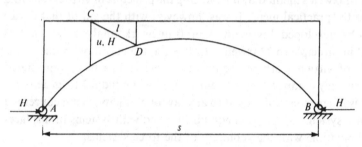

Fig. 42. From Mohr (1874*a*).

simplified by virtue of symmetry and superposition, as shown in Fig. 41. Having dealt with the preliminaries in detail, Mohr introduced the principle of virtual work. Thus, with reference to Fig. 42 (his Fig. 21), if *CD* is the only bar capable of deformation and if the force in that bar caused by an abutment thrust *H* is *uH*, by using the compatible elastic displacements Δs between the abutments and Δl of *CD*:

$$\left. \begin{aligned} -H\Delta s &= uH\Delta l \\ -\Delta s &= u\Delta l \end{aligned} \right\}$$

or   [(5.19)]

to provide a relationship whereby the small change of length of any bar may be related to a consequential small change in the span *s*. This relationship is used in the equation which specifies that Δs is zero when the abutments are rigid; and thereby, the value of *z* which is the horizontal thrust due to a load $P = 1$, is obtained as follows:

$$\left. \begin{aligned} 0 &= 2z \sum_{0}^{s/2} ru^2 + \sum_{0}^{a} rvu + \sum_{a}^{s/2} rwu \\ z &= -\left\{ \sum_{0}^{a} rvu + \sum_{a}^{s/2} rwu \right\} \Big/ 2 \sum_{0}^{s/2} ru^2 \end{aligned} \right\}$$

or   [(5.20)] where $r = l/EF$; and v and w are the forces in the bars when $P = 1$ (v being for those bars between $x = 0$ and $x = a$, and w for those bars between $x = a$ and $x = l/2$, while F is bar cross section).

In the first part of the second of Mohr's articles on the analysis of statically-indeterminate structures (1874b) he deals with bar frameworks with more than one redundant (supernumerary) bar, with regard to the requirements for equilibrium. For the sake of illustrating the nature of the problem he uses the framework shown in Fig. 43. He denotes the forces

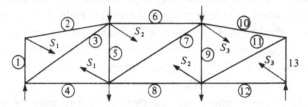

Fig. 43. From Mohr (1874b).

in the bars of the framework without the redundant bars shown, by U_1 when unit tension is applied in the line formerly occupied by redundant bar 1; and by U_2 when unit tension is applied instead in the line formerly occupied by redundant bar 2; similarly for U_3 (and any other redundant bars in general). Then, using r to denote the extension of a bar, due to unit tension, he uses the principle of virtual work to express the deflexions, due to the unit tensions in the lines formerly occupied by the redundants, as:

line 1: $\Sigma U_1{}^2 r$; line 2: $\Sigma U_2 U_1 r$; line 3: $\Sigma U_3 U_1 r$

for unit tension in line 1. Then:

line 1: $\Sigma U_1 U_2 r$; line 2: $\Sigma U_2{}^2 r$; line 3: $\Sigma U_3 U_2 r$

for unit tension in line 2 and:

line 1: $\Sigma U_1 U_3 r$; line 2: $\Sigma U_2 U_3 r$; line 3: $\Sigma U_3{}^2 r$

for unit tension in line 3.

For the changes in length of lines 1, 2 and 3 respectively, due to the external loading, Mohr writes by virtual work, $\Sigma U_1 Gr$, $\Sigma U_2 Gr$ and $\Sigma U_3 Gr$, where G is the force in a bar caused by the loading. Then the equations for the compatibility of the strains (deflexions and changes in length) of the statically-determinate system and the redundant bars are:

$$\left.\begin{array}{l}\text{(line 1) } 0 = \Sigma U_1 Gr + S_1 \Sigma U_1{}^2 r + S_2 \Sigma U_1 U_2 r + S_3 \Sigma U_1 U_3 r + ..\\[4pt] \text{(line 2) } 0 = \Sigma U_2 Gr + S_1 \Sigma U_2 U_1 r + S_2 \Sigma U_2{}^2 r + S_3 \Sigma U_2 U_3 r + ...\\[4pt] \text{(line 3) } 0 = \Sigma U_3 Gr + S_1 \Sigma U_3 U_1 r + S_2 \Sigma U_3 U_2 r + S_3 \Sigma U_3{}^2 r + ...\end{array}\right\} [(5.21)]$$

(where S_1, S_2, S_3 are the forces in the redundants and the summations $\Sigma U_1{}^2 r$, $\Sigma U_2{}^2 r$ and $\Sigma U_3{}^2 r$ include the respective redundant bar).

Mohr proceeds to deal with the inclusion of temperature effects and concludes by considering certain popular forms of a symmetrical bridge girder (truss). He includes the cross-braced girder, statically-indeterminate because of the cross-bracing of the quadrilateral panels, and notes the expedient used in practice whereby, for symmetrical loading, the structure is considered as consisting of two statically-determinate forms. Both contain the booms and vertical bars but they differ in that one has a set of diagonal bars which is different from the other. Half the loading is considered as applied to each structure at the same points and the two structures are analysed by statics alone, the results being combined to give the forces in the bars of the complete structure. He discusses the approximation implicit in the method.

Typical of later development of the method is the analysis of a continuous bridge truss of the Warren type shown in Fig. 44 (Mohr's

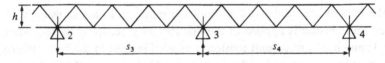

Fig. 44. From Mohr (1875).

Fig. 20) of the second part of his second article 'Beitrag zur Theorie des Fachwerks' (1874b, 1875). Choosing the statically-indeterminate quantities (forces in the elements supernumerary to the requirements of statics) as the forces in the bars connecting the spans over each intermediate support, he gives the first three (of n if there are $n+1$ spans) equations of compatibility of strain relating to those bars as:

$$(18) \quad \left. \begin{array}{l} 0 = \Sigma^{1,\,2} U_1 \, Gr + S_1 \Sigma^{1,\,2} \ U_1{}^2 r + S_2 \Sigma^2 U_1 \, U_2 r \\ 0 = \Sigma^{2,\,3} U_2 \, Gr + S_1 \Sigma^2 U_2 \, U_1 r + S_2 \Sigma^{2,\,3} U_2{}^2 r + S_3 \Sigma^3 U_2 \, U_3 r \\ 0 = \Sigma^{3,\,4} U_3 \, Gr + S_2 \Sigma^3 U_3 \, U_2 r + S_3 \Sigma^{3,\,4} U_3{}^2 r + S_4 \Sigma^4 U_3 \, U_4 r \end{array} \right\} \quad [(5.22)]$$

where the first terms, obtained by virtual work represent deflexions in the lines 1, 2 and 3 respectively, caused by the loading applied to the structure without bars in those lines; while the other terms relate to the effects of the forces S_1, S_2 and S_3 (statically-indeterminate quantities) in bars occupying the lines 1, 2 and 3.

Further details are apparent from Mohr's derivation (in a footnote) of the analogy between these equations and those of Clapeyron's (theorem of the three moments, Chapter 2) for a uniform continuous beam, which Mohr illustrates with reference to the second of his equations (18) above and which he finally puts in the form:

$$M_1 s_2 + M_2(s_2 + s_3) + M_3 s_3 = \tfrac{1}{4}(q_2 s_2{}^3 + q_3 s_3{}^3). \qquad [(5.23)]$$

where s_2 and s_3 are the respective spans and q_2 and q_3 the intensities of distributed load, with couples M_1, M_2 and M_3 over (redundant) supports 1, 2 and 3 respectively, in accordance with Clapeyron's theorem for uniform continuous beams.

The remainder of Mohr's article, from which this example is taken, is devoted to notable devices for facilitating the calculation of deflexion of frameworks, especially bridge trusses, by graphical and analytical methods (which are considered in Chapter 10).

Levy, 1874

Levy's contribution (1874), simultaneous with that of Mohr, is quite different. Because it is part of a broad study of the nature and economy of bar frameworks, which is unique in some important respects and little known, it is considered in detail in Chapter 6. But the following example (shown in Fig. 45) is sufficient to give the essence of his method

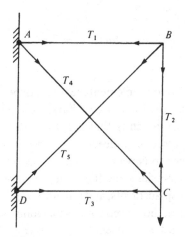

Fig. 45

of analysing statically-indeterminate frameworks: it has one redundant bar or member which is supernumerary to the requirements of statics. If the members of the plane, pin-jointed framework shown have an elastic extension (or compression) per unit length of a then, assuming that all such elastic changes in length caused by the load F are small, so that there is negligible change in geometry of the structure:

$$\left.\begin{array}{l} AC^2 = AB^2 + BC^2 \\[2mm] AC^2 = AD^2 + CD^2 \end{array}\right\} \tag{5.24}$$

that is, if the small extensions of the bars are denoted by $e_1, e_2, \ldots$

$$\left.\begin{array}{l} (2l+e_4)^2 = (l+e_1)^2 + (l+e_2)^2 \\[2mm] (2l+e_4)^2 = (l+0)^2 + (l+e_3)^2 \end{array}\right\} \tag{5.25}$$

Also:

$$\left.\begin{array}{l} BD^2 = BC^2 + CD^2 \\[2mm] BD^2 = AB^2 + AD^2 \end{array}\right\} \tag{5.26}$$

or

$$\left.\begin{array}{l} (2l+e_5)^2 = (l+e_2)^2 + (l+e_3)^2 \\[2mm] (2l+e_5)^2 = (l+e_1)^2 + (l+0)^2 \end{array}\right\} \tag{5.27}$$

After adding the four relationships obtained above and simplifying, including the elimination of second-order small quantities, the following equation of compatibility of strains emerges:

$$2l(e_4 + e_5) = l(e_1 + e_3) + l(e_2 + 0) \tag{5.28}$$

and since $e_1 = alT_1$; $e_2 = alT_2$; $e_3 = alT_3$; $e_4 = 2alT_4$; $e_5 = 2alT_5$:

$$2(T_4 + T_5) = (T_1 + T_3) + T_2 \tag{5.29}$$

which is the additional equation needed to supplement the four independent equations afforded by statics to relate the forces in the bars and the load.

A noteworthy feature of Levy's approach is that it is unnecessary to identify the redundant bar at the outset or, indeed, at any stage of the analysis. It bears a strong resemblance to Navier's method in the sense of conceptual simplicity. But whereas Navier's method deals in terms of resolution of both forces and elastic displacements, and is readily applicable to any kind of bar framework, Levy's method, even with its unsophisticated use of statics and geometry, does not lend itself to the same ease of application in all circumstances. Since Levy refers to Navier, it is likely that he was attempting to improve on the latter's exposition and, at the same time, arrange the analysis in terms of bar forces, tensions and loads, rather than displacements of joints. It is noteworthy that while the methods of Navier (and Clebsch) established a particular general approach to structural analysis (which is independent of the type of structure), whereby equations of equilibrium of joints are formulated finally in terms of deflexions of joints, the method of Maxwell and Mohr, on the other hand, results in final equations of compatibility of strain in terms of the forces in the redundants (supernumerary elements). The former has been described as the 'equilibrium approach' (or the 'displacement method') for the sake of brevity (though both descriptions are clearly inadequate): the latter has been described (also inadequately) as the 'compatibility approach' (or the 'force method'). Each approach is represented in an energy principle, as described in Chapter 8. But Levy's method belongs to neither of the two systematic approaches: it possesses the feature, however, that no choice is necessary *a priori* with regard to the designation of redundants, a commendable feature in common with Navier's method.

Mohr's claim to priority

It is revealing here to recall Mohr's discussion of the origin of his method of analysing statically-indeterminate frameworks. In his article (1885) he claimed that the short but important article, 'On the calculation of the equilibrium and stiffness of frames',

which Professor Maxwell published in the *Philosophical Magazine* as early as 1864, has been unknown in Germany until recently...Since that journal is not received in the library at Dresden, I first became aware of the article through a paper by Swain, 'On the application of the principle of virtual velocities to the determination of the deflection and stresses of frames' in the *Journal of the Franklin Institute* for 1883.

Mohr noted that Maxwell had not used the principle of virtual velocities. Rather, Mohr said, he used Clapeyron's theorem in order to prove the principle of reciprocity of deflexions, first for the determinate and then for the statically-indeterminate framework; thus, by this means, Maxwell had determined the forces in the supernumerary bars when the forces in the framework are due to external forces only, and that he had obtained this in a similar form to that which Mohr himself proposed (1874*a*, 1874*b*).

Mohr seemed to be unaware of Jenkin's priority (1869) with regard to the use of virtual work instead of Clapeyron's theorem, to obtain Maxwell's equations. He must, therefore, have ignored Swain's reference (1883) to Jenkin. It is, moreover, a little surprising that Mohr did not comment on Swain's blatant misuse (or even misunderstanding) of the description 'principle of virtual velocities' in describing his method which involved merely the direct use of the law of conservation of energy (manifest in Clapeyron's theorem!). Mohr's attitude to other contemporaries is considered in Chapter 10. But Mohr must have been gratified by the acknowledgement of the distinguished Regius Professor Winkler of Berlin. Having discussed principles using Levy's approach in his celebrated theory of bridges (1881b, vol. 2), Winkler proceeded to commend Mohr's novel method for statically-indeterminate frameworks. (Indeed, Winkler makes liberal references, throughout his exhaustive treatise, to analytical devices by Mohr). It is noteworthy that Maxwell (1864b) is included in Winkler's (1881b) bibliography, but Winkler seemed unaware of Castigliano (1879).

The reciprocal theorem

An important property of linearly elastic structures, which emerged from the researches of Clebsch, Maxwell and Mohr, is the reciprocal theorem. It was apparently also discovered, independently, by Betti and Rayleigh (the former used the principle of virtual work in an approach which is especially interesting with regard to theory of structures). It is interesting that opinions seem to have varied, however, regarding the origin of the theorem. For example, Love, the distinguished British elastician attributed (1927) the theorem to Betti (1872) but said that it is embodied in a more general theorem due to Lord Rayleigh (1873). Rayleigh, however, acknowledged (1894) Betti as the originator of the reciprocal theorem. Neither Rayleigh nor Love mentioned Maxwell's version (1864b). Also, although Rayleigh and Love quoted Clebsch's contributions to theory of elasticity from his celebrated book (1862), that part of his work (Chapter 8 on systems of bars) which exhibits the reciprocal property of linearly elastic systems, apparently escaped their attention. For some time Mohr believed that credit for the observation of the reciprocal property of linear structures belonged to him (for he first published an article about it in 1868), but eventually he recognised Betti (to judge, for example, from the contents of his book (1906) of collected topics in technical mechanics).

It is appropriate and interesting to recall the nature of Betti's derivation of his theorem which described the reciprocal property of linearly elastic systems. If a system consists of an assemblage of elastic bars pin-jointed together to form a load-bearing structure, and if loads $F_1, F_2, \ldots, F_n$ are in equilibrium with forces $T_1, T_2, \ldots, T_N$ in the bars, and small deflections $\Delta_1, \Delta_2, \ldots, \Delta_n$ occur in the lines of action of those loads, which are compatible with small changes in length $e_1, e_2, \ldots, e_N$ of the bars (directly proportional to the forces T), then a different condition of loading $F_1', F_2' \ldots, F_n'$ is associated with bar forces $T_1', T_2', \ldots, T_N'$ and compatible small elastic displacements $\Delta_1', \Delta_2', \ldots, \Delta_n'$ and $e_1', e_2', \ldots, e_N'$. It is assumed that n represents the total number of points at which loads may be applied and that loads may be applied in any direction at any point. Now, for the purpose of applying the principle of virtual work to the two specified systems of forces in equilibrium, Betti observed that either set of compatible displacements associated with those force systems may be used as virtual displacements. Thus, by the principle of virtual work:

$$\sum_{i}^{n} F_i \Delta_i' = \sum_{j}^{N} T_j e_j'$$ (5.30)

or

$$\sum_{i}^{n} F_i' \Delta_i = \sum_{j}^{N} T_j' e_j$$ (5.31)

provided that Δ_i' is the component in the line of action of F_i and Δ_i is the component in the line of action of F_i'. Since $e_j = a_j T_j$ and $e_j' = a_j T_j'$, where a_j is the elastic coefficient of the jth bar:

$$\sum_{i}^{n} F_i \Delta_i' = \sum_{j}^{N} a_j T_j T_j'$$ (5.32)

and

$$\sum_{i}^{n} F_i' \Delta_i = \sum_{j}^{N} a_j T_j T_j'$$ (5.33)

whence

$$\sum_{i}^{n} F_i \Delta_i' = \sum_{i}^{n} F_i' \Delta_i$$ (5.34)

Betti's own proof was more sophisticated and referred to an elastic solid (see Notes). The physical significance of the result is not immediately apparent and it is hardly surprising that the simple exposition of Maxwell, some eight years earlier, has earned him widespread credit, such that Maxwell's reciprocal theorem is the usual manner of its description.

Influence lines for deflexion

Mohr, it seems, was aware of the reciprocal property of linearly elastic systems almost as soon as were Clebsch and Maxwell. In an article (1868) he considered the deflexion of a uniform simply supported beam whose material obeyed Hooke's law and showed that its deflexion, due to a concentrated vertical load P_0 at any point C, is such that the deflexion at any other point B is the same as the deflexion at C when the load is applied at B. He concluded, moreover, that the curve of elastic deflexion, due to the load at C, represents the variation of deflexion at C as a concentrated load traverses the beam. Indeed, he described the curve as representing the *influence line* for the deflexion of C. Fig. 46 shows Mohr's illustration of the principle in his collection of topics on technical

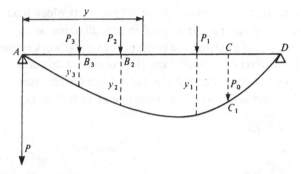

Fig. 46. After Mohr (1875).

mechanics (1914, 2nd edn, chapter 10). When a train of loads P_1, P_2, P_3 is in the position shown, the deflexion of C is:

$$y = \frac{1}{P_0}(P_1 y_1 + P_2 y_2 + P_3 y_3) \qquad [(5.35)]$$

That is, the curve AC_1D is the influence line as specified.

Finally, it is appropriate to recall the article by Professor Krohn of the Aachen Polytechnikum under the title of 'The law of reciprocal displacements and its use in analysing statically-indeterminate frameworks' (1884). There he demonstrates the principle, first by using displacement diagrams to calculate deflexions of a simple bar framework, then he proceeds to use it for analysing a two-pin framed arch where the abutment thrust represents the single degree of statical indeterminacy. Reference is made to the work of Williot (1877*a*, 1877*b*) with regard to displacement diagrams.

Notes

Under the title 'Teorema generale intorno alle deformazioni che fanno equilibro a forze che agiscono soltanto alle superficie' (1872), E. Betti derived the following equation of virtual work for a linearly elastic body:

$$\int_\sigma (L'u'' + M'v'' + N'w'')\,d\sigma = \int_\sigma (L''u' + M''v' + N''w')\,d\sigma$$

with regard to applied forces L', M', N' causing elastic deflexions in their lines of action of u', v', w' and, alternatively, L'', M'', N'' with deflexions u'', v'', w''. He then gave the theorem (which, he said, was a consequence of that equation) as follows:

If a homogeneous elastic solid body is in equilibrium for two separate systems of applied forces, the sum of the products of the components of the first system of forces and the components of deflexion of the second system of forces is equal to the sum of the products of the components of the second system of forces and the components of deflexion of the first.

Müller-Breslau (1887*b*: vol. 2, 1892) remarks with reference to Mohr's article (1874*b*: part 3, 1875), that 'here for the first time the elastic line of a truss is treated as a link polygon'.

It is noteworthy that Mohr (1875) treats continuous trussed girders as statically-indeterminate frameworks, whereas, in his earlier work on continuous beams (1860–8), he avoided the difficulty by using the concept of equivalent solid sections (Chapter 2).

6

Levy's theory of frameworks and bridge girders

In his book *La statique graphique* of 1874, Maurice Levy, Ingénieur des Ponts et Chaussées and former pupil of Saint-Venant, provides a detailed study of the equilibrium, stiffness and economy of bar frameworks and girders, which merits particular attention, especially because it seems to have been generally overlooked. It is contained in note 2 of the book, which covers some ninety pages of analysis and results and which is largely unrelated to the title of the book as a whole or to its other contents.

He is especially concerned with demonstrating that statical indeterminacy within a framework affords no advantage with regard to stiffness or economy in respect of quantity of material. But it is implicitly acknowledged, for example by his Fig. 120 (shown in Fig. 47) that continuity over

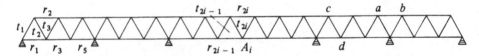

Fig. 47. From Levy (1874).

numerous supports of an otherwise statically-determinate bar-frame girder is advantageous in certain circumstances.

Having acknowledged Navier and others with regard to the general analysis of statically-indeterminate frameworks and structures, Levy gives a theory and method of analysing such frameworks, which is essentially a sophisticated and instructive inversion of Navier's original method.

In the introduction to his book (1874, para. 10), Levy writes:

In note 2 we expound a very simple general method for determining, by means of mathematical theory of elasticity, the tensions in the bars of an elastic system which is indeterminate by statics alone.

It is a *résumé* of a memoir which we have recently presented to the Academy of Sciences (*Comptes rendus des séances de l'Académie des Sciences*, 28 April 1873). It is written in a manner which makes it independent of the rest of the book. We believe it contains some novel results, interesting by themselves and useful in structural mechanics. One discovery is that those systems of bars which are amenable to graphical statics can be constructed so that bars in either tension or compression carry the same force per unit area of section. They are then the most economical and desirable and show that our researches in this field, undertaken with the aim of exposing, by the mathematical theory of elasticity, the shortcomings of graphical statics, have led to an unforeseen feature, a new and important property of this latter science.

Then the beginning of note 2 of the book (1874, pp. 236–9) reads as follows:

Memoir on research into forces in the bars of elastic systems and those systems which, for the same (or a given) volume of material, provide the greatest possible stiffness. Objective and Main Results of this Memoir.

Systems of bars or elastic lines will be considered whose joints are either pinned or spherical: pinned if the bars are all in one plane; spherical otherwise.

Forces are supposed to be applied externally to the system at the joints. It is proposed to find methods whereby the forces in the bars of a system may be calculated.

Sometimes statics alone is sufficient to solve this problem: but frequently it is insufficient and, for the purpose of finding solutions, it is necessary in addition to appeal to the elastic properties of the material or the materials of which the bars are made.

Later, Levy finds that, for a system of bars arranged in any way in one plane or in space, in equilibrium under the influence of given forces, a condition which must be satisfied in order that the forces in the bars may be found by statics alone, is that the geometrical figure formed by the bars must not contain supernumerary lines, that is, no more lines than are necessary to define the overall form of the system.

Having suggested that the most important problem, from the viewpoint of practical application, consists in determining the cross-sections of a system of bars forming a given shape and subjected to specified forces; then, in order that all the bars in tension bear the same stress and all the bars in compression bear the same stress, he proposes:

1. In order that a system consisting of *m* bars in equilibrium under the action of specified forces, may be constructed as a solid of uniform resistance or strength, it is generally necessary and *always* sufficient that the shape formed by the axes of the bars does not contain supernumerary lines.

2. The number of ways that a shape formed by *m* lines and containing

k supernumerary lines can be made into a solid of uniform strength is k-fold, that is to say, one can in this case choose arbitrarily (between certain limits) the sections of k-bars and, provided that the sections of the $m-k$ remaining bars are suitably chosen, all of the bars perform the same duty.

On the basis of these propositions it is concluded that the best constructions, those which use the least material, are, in general, the most simple.

'This conclusion gives, we suggest, a new importance to graphical statics, whereby simple procedures and devices replace so advantageously calculations in ordinary statics which are laborious and often intractable.' He believes that, in the U.S.A., major bridges which use the principle of articulation are designed on this basis.

Levy goes on (1874, p. 239) to discuss the possibility of a framework of uniform strength with supernumerary bars, which requires no more material for the same specified loading than a framework for the same duty without supernumerary bars. He discusses the consequential variety of ways of arranging supernumerary bars and their dimensions, and proposes the theorem:

When a system of bars contains supernumerary members and satisfies the conditions which make it possible in one way (and, therefore, in a variety of ways) to make it a body of uniform strength, a system necessarily exists which consists of some of those bars, but without supernumerary members, being a system of uniform strength and having the same volume of material as the original system.

In a footnote Levy considers circumstances where supernumerary members are essential, thus:

We do not pretend that in practice supernumerary members can always be eliminated; but we suggest the following approach which we believe to be both sensible and scientific: the basis of a structure in wood or metal may be a configuration without supernumerary members whose dimensions are determined strictly by theoretical principles and such that the interaction of the bars is determined by the joints between them only. If, however, among those elements in compression there are some which are so long that they are liable to bend out of line under their working load, it becomes necessary to restrain them by additional elements which, though occupying supernumerary lines, may be of very small cross-section, being merely to stabilise main members [bars] and not to contribute to the stiffness of the structure of themselves. It is emphasised that it is essential that such secondary elements are not made of substantial proportions: for, if they are, the force distribution within the framework can be altered and the dimensions, calculated originally for the principal elements, may be invalidated.

A remarkable work where, we believe, these precepts have been observed, is the Pont d'Arcole. The arch is taken to the crown by simple triangles. Only the very long sides of these triangles are restrained by very slender secondary members; and it is precisely because this work is designed in such a rational way that it combines great strength with great elegance.

Levy urges simple configurations for structures and says that a simple triangulated girder is preferable to one which takes the form of the cross of St Andrew but this latter, in its turn, is preferable to a lattice girder, and so on. He suggests that to explore the configurations to support specified loads, using the least volume of material for uniform strength, it is only necessary to consider those without supernumerary bars. This he has done with particular reference to the configurations of girders in common use in Europe and the U.S.A., with the help of his friend Brune, former pupil of l'Ecole Polytechnique and l'Ecole des Beaux-Arts and professor at the latter. The results are tabulated to provide comparison of the volumes of material necessary for resisting dead load, live load and combinations of those loads for various kinds of triangulated girders. He observes that, generally, the newer American girders, though of attractive appearance, are unlikely to be the most economical and that the most simple triangular forms, little known in France, but for a long time common in England by the name of Warren girders, are distinctly preferable. He concludes that, in principle, simply-triangulated girders are best and accord with the precepts of the theory; the tabulated data obtained from his calculations show that the economies afforded are not trivial but very substantial (as noted below).

After thus explaining the objective of note 2, Levy proceeds to state and justify certain theorems; first 1, 2 and 3 (pp. 243–7) and subsequently, theorems 4, 5 and 6 (pp. 255–76). They are as follows:

Theorem 1

In order that statics may be able to furnish the forces in a system of bars, it is necessary and sufficient for the geometrical figure formed by the axes of the bars to be such that it may be constructed by giving, arbitrarily, the lengths of all its sides. (It is noteworthy that Levy uses the principle of virtual velocities concerning this theorem and discusses that principle at some length.)

Theorem 2

Whenever the geometrical figure formed by a system of bars contains k supernumerary lines, statics furnishes k equations fewer than necessary to define the forces in the bars and, inversely, if statics furnishes

k equations too few to define the forces in a system of bars, one may be sure that the geometrical figure which they form contains k supernumerary lines.

Theorem 3

If it is possible to construct, by taking arbitrary lengths of all its sides, a figure formed by m bars, without violating the conditions of connection of the bars, then statics will enable the forces in all the bars to be found, together with the reactions of their supports. But if, in respect of the conditions whereby the joints are determined, knowledge of $m-k$ of m sides of the figure is sufficient, then statics provides k equations too few for finding the tensions in the bars and the reactions of the supports.

Theorem 4

In order that a system of bars in equilibrium under the action of given forces may be regarded as a body of uniform strength, it is necessary in general, and *always* sufficient that it contains no supernumerary lines.

In an exceptional case in which a system contains supernumerary lines, achievement of uniform strength is possible in an infinite number of ways, that is to say there are an infinite number of ways of modifying the sections of the bars to achieve uniform strength.

Lemma

If a figure with k supernumerary lines is so constructed that it is possible, in one way (or, consequently, in an infinite number of ways), to arrange the sections of the bars so that a system of uniform strength is obtained relative to given external forces, one can always, by leaving out some of the bars, form a system of uniform strength without supernumerary lines, which, with respect to the same forces, deforms elastically in a manner identical to the original system.

Theorem 5

When a system containing k supernumerary lines and made of bars of the same material is such that it is possible, in any one way (or, consequently, in an infinite number of ways), to be made into a system of uniform strength with regard to specified external forces, then there is always a system, without supernumerary bars, capable of resisting the same forces, with the same amount of material.

Theorem 6

When a system, containing k supernumerary lines and such that it can, in one way (and, in consequence, in an infinite number of ways), be a system of uniform strength, relative to specified forces acting upon it, there always exists a system, without supernumerary lines, capable of resisting the same forces and such that the sum of the products of volumes of bars and their coefficients of elasticity respectively, is the same for that system and the given system.

Theorem 5 may be justified (with reference to the Lemma preceding) on the basis of a structure of uniform strength which, by definition, regardless of the presence of supernumerary bars, must always be of the same stiffness with respect to a specified function and loading if the limiting stress is invariable (and, by definition of uniform strength, exists in every part of the structure).

Using Levy's notation, if ϵ is the limiting value of t/we throughout, where t is the tensile or compressive force in a bar and we the product of cross-sectional area and modulus of elasticity, then the strain energy T of the structure, being equal to the work done by the specified loading, is:

$$T = \frac{\epsilon^2}{2} \Sigma a_i \omega_i e_i \qquad [(6.1)]$$

the summation including all bars, where a_i is the length of a bar. Then if e is the same for the whole structure:

$$\left. \begin{aligned} T &= \frac{e\epsilon^2}{2} \Sigma a_i \omega_i \\[2mm] \Sigma a_i \omega_i &= \frac{2T}{e\epsilon^2} \end{aligned} \right\}$$

or   (Levy's equation 31)   [(6.2)]

Therefore, since $\Sigma a_i \omega_i$ is the volume of material in the structure and since $2T/e\epsilon^2$ is the same for all structures of uniform strength which fulfil the specified conditions, the theorem is verified.

After theorem 3 there is section 2 of note 2 headed 'General method for finding the tensions in a system of elastic bars when statics leaves the problem unsolved'. Having briefly discussed the value of the principle of virtual velocities with regard to the conditions of equilibrium within bar frameworks (during which he specifically refers to compatible virtual elongations of the bars), he proceeds to define k geometrical relationships which must be used if these are in elastic bars, but only $m - k$ equations of statics for the system. Thus, he postulates that there are k relationships of the kind:

$$F(a_1+\alpha_1, a_2+\alpha_2, \ldots, a_m+\alpha_m) = 0 \left.\vphantom{\begin{matrix}1\\1\\1\end{matrix}}\right\}$$

where $\qquad\qquad\qquad\qquad\qquad\qquad\qquad\qquad\qquad\qquad\qquad$ [(6.3)]

$$\alpha_i = a_i t_i / e_i \omega_i$$

if t_i is the force (or tension); ω_i the section; and e_i the coefficient of elasticity of the ith bar.

In order to make those relationships linear it is noted that $\alpha_1 = \alpha_2 = \ldots = \alpha_m$ must be zero when

$$F(a_1, a_2, \ldots, a_m) = 0 \qquad\qquad\qquad\qquad\qquad\qquad [(6.4)]$$

and that each of k such relationships must be valid also when α_i is small, in accordance with the 'usual assumptions in the mathematical theory of elasticity'.

If, for the sake of brevity, $F(a_1, a_2, \ldots, a_m)$ is denoted simply by F, then:

$$\frac{dF}{da_1}\alpha_1 + \frac{dF}{da_2}\alpha_2 + \ldots + \frac{dF}{da_m}\alpha_m = 0 \qquad\qquad [(6.5)]$$

(that is, if $F = 0$ then $\delta F = 0$ because $\alpha_1 = \delta a_1, \ldots$) and, since $\alpha_i = a_i t_i / e_i \omega_i$,

$$\frac{dF}{da_1}a_1\frac{t_1}{e_i\omega_i} + \frac{dF}{da_2}\alpha_2\frac{t_2}{e_2\omega_2} + \ldots + \frac{dF}{da_m}a_m\frac{t_m}{e_m\omega_m} = 0 \qquad [(6.6)]$$

followed by the remark 'such are the k relationships to supplement those furnished by statics to determine the bar forces t_i'. Then Levy notes as follows:

Remark 1. All the equations which furnish the 'tensions' being linear, one notes that, if two systems of forces are applied simultaneously, at different points of a framework, the tensions which they cause in the various bars are the same as those obtained respectively by adding the effect of each of the two systems of applied forces acting alone. (Thus, he states the principle of superposition.)

Remark 2. When statics by itself suffices to determine the tensions in a system of bars, those tensions depend only on the geometry of the figure formed by the axes of the bars: but when statics leaves some tensions undetermined and where, for the purpose of evaluating them, it is necessary to appeal to elastic properties, the tensions depend not only on the geometry of the system but also on the sections ω_i and the coefficients of elasticity e_i of the bars. In the first instance, if the sections of the bars are changed then their tensions remain unchanged, while only their elongations vary; in the second instance it is possible in an infinity of ways to alter the tensions of the various bars by changing their sections appropriately.

Levy demonstrates the application of his method to the plane framework shown in Fig. 48 (his Fig. 116 where points a_0, a_1, a_2, a_3 are distant a apart respectively) for the loading shown. There are clearly four bar tensions to be found and since there are only two independent equations of statics,

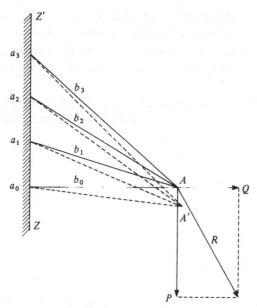

Fig. 48. From Levy (1874).

two equations must be obtained by appeal to the geometry of elastic deformation and the law of elasticity. Use of the theorem of Pythagoras enables these latter to be obtained, for example:

$$-b_0^2+b_2^2 = 4a^2 \quad \text{(i)}$$
$$b_1^2-b_0^2 = a^2 \quad \text{(ii)}$$

[(6.7)]

Multiplying (ii) by two and subtracting from (i) gives:

$$b_0^2+b_2^2 = 2b_1^2+2a^2$$

[(6.8)]

which Levy quoted without proof in the general form:

$$b_i^2+b_{i+2}^2 = 2b_{i+1}^2+2a^2$$

[(6.9)]

which is relevant when the system has any number of bars. He denotes the (small) elongation of the ith bar by β_i and writes:

$$(b_i+\beta_i)^2+(b_{i+2}+\beta_{i+2})^2 = 2(b_{i+1}+\beta_{i+1})^2+2a^2$$

[(6.10)]

whence, if second-order small quantities are neglected,

$$b_i\beta_i+b_{i+2}\beta_{i+2} = 2b_{i+1}\beta_{i+1}$$

[(6.11)]

and, since $\beta_i = b_i t_i/e_i\omega_i$,

$$b_i^2\frac{t_i}{e_i\omega_i}+b_{i+2}^2\frac{t_{i+2}}{e_{i+2}\omega_{i+2}} = 2b_{i+1}^2\frac{t_{i+1}}{e_{i+1}\omega_{i+1}}$$

[(6.12)]

whence, 'The product of the tension of a bar per unit section and unit coefficient of elasticity, multiplied by its length, is the mean of the similar products of the two adjacent bars.' He then notes that the (two) additional equations required to solve the kind of framework shown are thus available.

Levy's second example uses the framework shown in Fig. 49 (his Fig. 115), a framework formed by a rectangle and its two diagonals. Again he

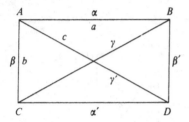

Fig. 49. From Levy (1874).

does not solve the problem completely, concentrating only on the derivation of the condition of compatibility of strain which is required in this instance to supplement the five independent equations of statics relating the tensions and external forces which are applied externally at the corners A, B, C and D of the framework to maintain it in equilibrium.

Levy specifies forces f_1, f_2, f_3, f_4 acting respectively at the four corners of the rectangular framework and in its plane, and notes that a total of eight equations of statics may be obtained regarding the equilibrium of those joints, though because the four applied forces must satisfy three equations of equilibrium of the framework as a solid body, only five independent equations remain to determine the tensions in the bars. He then states that since there are six unknown tensions, there is evidently a geometric relationship between the elastic elongations. That relationship, which he suggests is 'very easy' to obtain, is:

$$a(\alpha+\alpha_1)+b(\beta+\beta_1) = c(\gamma+\gamma_1) \qquad [(6.13)]$$

where a, b, c are the original lengths of the pairs of similar bars (c being the length of the diagonals), and α and α_1; β and β_1; γ and γ_1 their respective elongations. It is, however, believed to be worthy of proof, as follows:

$$\left.\begin{aligned}
(a+\alpha)^2+(b+\beta)^2 &= (c+\gamma)^2 \\
(a+\alpha_1)^2+(b+\beta_1)^2 &= (c+\gamma_1)^2 \\
(a+\alpha_1)^2+(b+\beta)^2 &= (c+\gamma)^2 \\
(a+\alpha)^2+(b+\beta_1)^2 &= (c+\gamma_1)^2
\end{aligned}\right\} \qquad (6.14)$$

being four possible equations, by the theorem of Pythagoras, if the elongations are small and insufficient to change the essential geometry of the framework. Addition of these equations gives:

$$2(a+\alpha)^2 + 2(a+\alpha_1)^2 + 2(b+\beta)^2 + 2(b+\beta_1)^2$$
$$= 2(c+\gamma)^2 + 2(c+\gamma_1)^2 \tag{6.15}$$

whence, neglecting small quantities of the second order,

$$a(\alpha+\alpha_1) + b(\beta+\beta_1) = c(\gamma+\gamma_1) \tag{6.16}$$

Substitution of, say,

$$\alpha = a\frac{t_1}{e_1\omega_1}; \quad \alpha_1 = a\frac{t_2}{e_2\omega_2}; \quad \beta = b\frac{t_3}{e_3\omega_3}; \quad \beta_1 = b\frac{t_4}{e_4\omega_4};$$

$$\gamma = c\frac{t_5}{e_6\omega_6}; \quad \gamma_1 = c\frac{t_6}{e_5\omega_5}$$

provides the required sixth relationship between the tensions.

Next Levy briefly discusses the analysis of the frameworks shown in Figs. 50 and 51 (his Figs. 117 and 121). Concerning the former he notes that statics alone enables the reactions of the supports and the tensions in the four end bars to be found, after which the method described above may

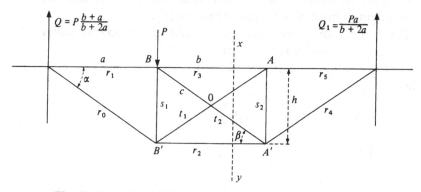

Fig. 50. From Levy (1874).

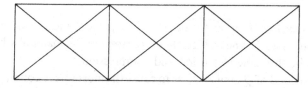

Fig. 51. From Levy (1874).

be used to enable the tensions in the other bars to be found. The latter, in which the panels are in the form of the cross of St Andrew, requires three equations of compatibility of strains (one for each of the three panels) to supplement the equations provided by statics and, again, the type of equation derived above is sufficient.

The final section (4) of note 2 of Levy's book, consisting of some forty-six pages is headed 'Comparison between the principal girders used in Europe and the U.S.A., from the point of view of the volume of material required to support dead or live loads'. The general conclusions are given in section 1, as described above, and are repeated at the beginning of section 4 by way of introduction. Then it is interesting that, later, Levy indicates that his study was influenced by a directive issued by the Minister of Public Works on 15 June 1869, in which an earlier requirement, whereby specified dead loading only was a sufficient criterion for the strength of bridges, was supplemented by one which specified criteria with regard to live loading as well. The reason, according to Levy, was an accident involving the failure of a bridge.

Also, in the introductory remarks, Levy is severely critical of American practice involving girders with supernumerary bars: he mentions those of Jones, Murphy–Whipple, Linville and Post and asserts that, owing to the inadequacy of statics alone for these structures, they are designed by semi-empirical methods which are often defective!

Finally, Levy summarises his investigations (which concerned four types of girder) and tables of results, as follows:

(a) For a uniformly distributed dead load only, the volumes of material are:

Fink $5.33\dfrac{PL}{R}$

Bollman $5.12\dfrac{PL}{R}$

Isosceles triangles $3.61\dfrac{PL}{R}$

Right-angled triangles $3.67\dfrac{PL}{R}$

if the ratio of depth h of girder to span L is $1/12$; P is the total load over the whole of the span L; and R is the limiting stress of the material which is assumed to be the same in tension and compression.

(b) For a live load of Q applied so as to cause the most severe conditions and a ratio of depth to span of $1/12$, the volumes of material for that load alone are:

Fink	$12.87\dfrac{QL}{R}$
Bollman	$163.84\dfrac{QL}{R}$
Isosceles triangles	$8.24\dfrac{QL}{R}$
Right-angled triangles	$8.48\dfrac{QL}{R}$

If, however, the proportions of each girder may be chosen for least volume of material (though of constant height throughout, or parallel chords for the triangulated girders) then the corresponding results are:

(c) Fink $\quad 2.30\dfrac{PL}{R}$, with $\delta=\dfrac{h}{L}=0.28$

Bollman $\quad 1.57\dfrac{PL}{R}$, with $\delta=\dfrac{h}{L}=0.42$

Isosceles triangles $\quad 2.36\dfrac{PL}{R}$, with $\delta=\dfrac{h}{L}=0.14$

Right-angled triangles $\quad 2.41\dfrac{PL}{R}$, with $\delta=\dfrac{h}{L}=0.15$

But Levy notes that the apparent advantage of the Bollman girder is purely theoretical because a ratio of depth to length of 42/100 is inadmissible for all except the shortest spans. Thus, the girder of isosceles triangles (the Warren girder) is again that of least volume of material and, with the smallest ratio of depth of girder to span:

(d) Fink $\quad 8.36\dfrac{QL}{R}$, with $\delta=\dfrac{h}{L}=0.13$

Bollman $\quad 25.12\dfrac{QL}{R}$, with $\delta=\dfrac{h}{L}=0.42$

Isosceles triangles $\quad 5.66\dfrac{QL}{R}$, with $\delta=\dfrac{h}{L}=0.13$

Right-angled triangles $\quad 5.86\dfrac{QL}{R}$, with $\delta=\dfrac{h}{L}=0.13$

Levy ends his book by confirming that (within the bounds of his study) the Warren girder, or to use his terminology 'the triangulated system', is superior from the viewpoint of economy to that of Fink and much more so to that of Bollman. Also, it is based on the general theory of systems involving the cross of St Andrew and latticework and, therefore, on those of Jones, Linville, Murphy–Whipple etc., being derivatives of the lattice girder or cross of St Andrew and having supernumerary bars. Thus, elementary statics is sufficient for designing the most economical girders and for that purpose Levy asserts that graphical statics is expeditious.

7

Early developments of energy principles relating to theory of structures

This chapter is concerned primarily with principles involving energy concepts, which were revived or formulated early in the nineteenth century within the science of statics and the theory of elastic structures. Principles relating to energy, which were implicitly available at the beginning of the century, included the principle of virtual work (known then as the principle of virtual velocities) and the supreme law of conservation of energy.

Early history

According to Dugas (1955), the use of the principle of virtual velocities can be traced to Jordanus of Nemore in the thirteenth century (and to Aristotle's law of powers of the fourth century B.C.). A revealing account of the principle is provided by Mach (1883) and it is remarkable that the principle, and its use, preceded explicit recognition of conservation of energy. Mach discusses the meaning of the terminology at some length: he ascribes it to John Bernoulli and notes that the word 'virtual' is used in the sense of something which is physically possible. Bernoulli's definition of *vitesse virtuelle* is said by Mach to be incorporated in Thomson & Tait's wording: 'If the point of application of a force be displaced through a small space, the resolved part of the displacement in the direction of the force has been called its virtual velocity.' It is believed that the truth of the principle was first noted explicitly by Stevinus, at the close of the sixteenth century, in connection with his research into the equilibrium of systems of pulleys. Galileo is said to have recognised the validity of the principle in studying the problem of the inclined plane, but it seems that the universality of the principle was first recognised by John Bernoulli and communicated to Varignon in his letter of 26 January 1717.

Mach described the various deductions of the principle, including that of Lagrange, and the consequential discovery by Maupertuis of the *loi de repos* concerning equilibrium, which (according to Mach) he communicated to the Paris Academy in 1740 and which was discussed in detail by Euler in 1751 in the proceedings of the Berlin Academy. Maupertuis seems to have observed that equilibrium implies maximum or minimum energy ('work') of a system and, therefore, originated implicitly the powerful principle of minimum potential energy for stable equilibrium, a principle which, surprisingly, seemed to escape the attention of engineers concerned with theory of structures in the last century (and which was implied in the Bernoulli–Euler principle of *vis viva potentialis* for elastic bending beams *c.* 1740). According to Mach, Maupertuis then enunciated a principle which he called the 'principle of least action' in 1747 and declared that it accorded eminently with the wisdom of the Creator. But Mach believed that the derivation of the principle was based on an unclear mingling of his ideas of *vis viva* (energy) and the principle of virtual velocities. He described Euler's approach to the principle of least action and its satisfactory outcome, having discarded the metaphysical aspect of the principle as enunciated by Maupertuis. The principle of least action is not, however, believed to be relevant to the present study; neither is the later principle of least constraint which Mach attributes to Gauss (1829).

The principle of *vis viva* is strongly related, it seems, to the important concept of conservation of energy, which took so long to emerge in explicit terms. According to Mach, Huygens was the first to apply the principle in the seventeenth century and he was followed by John and Daniel Bernoulli in the eighteenth century, while in 1847 Helmoltz published his interpretation of it. (Maxwell (1877) refers to Helmoltz's 'celebrated memoir on conservation of energy' while Tait (1868) regarded him as one of the most successful of the 'early promoters of the science of energy on legitimate principles', who based the whole subject on Newton's principles and who embraced electrical and chemical theory. But, as early as 1842, Mayer, it seems, had suggested the equivalence of mechanical work and heat.) It is appropriate here to recall Clerk Maxwell's 'History of the doctrine of energy' from his book *Matter and motion* (1877). There is also an interesting communication by Tait to the *Philosophical Magazine*, dated 13 December 1864, under the title of 'A note on the history of energy', which is in accordance with Maxwell's account and includes the statement:

The opinion of James Bernoulli on a question of this nature would undoubtedly be valuable, but he seems not to have noticed Newton's remark. But I must protest against allowing any weight to that of John Bernoulli who, while inferior to his brother as a mathematician, was so utterly ignorant of the principle in question as seriously to demonstrate the possibility of a perpetual motion, founded on the alternate mixing of two liquids and their separation by means of a filter.

Maxwell writes:

The scientific importance of giving a name to the quantity which we call kinetic energy seems to have been first recognised by Leibnitz, who gave to the product of the mass by the square of the velocity the name *vis viva*. This is twice the kinetic energy.

Newton, in the *Scholium to the laws of motion*, expresses the relation between the rate at which work is done by the external agent, and the rate at which it is given out, stored up, or transformed by any machine or other material system, in the following statement, which he makes in order to show the wide extent of the application of the third law of motion:

'If the action of the external agent is estimated by the product of its force into its velocity, and the reaction of the resistance in the same way by the product of the velocity of each part of the system into the resisting force arising from friction, cohesion, weight, and acceleration, the action and reaction will be equal to each other, whatever be the nature and motion of the system.' That this statement of Newton's implicitly contains nearly the whole doctrine of energy was first pointed out by Thomson & Tait [Thomson & Tait, 1879, *Treatise on natural philosophy*, vol. 1, para. 268: in their preface they referred to the *grand principle of the conservation of energy*]. The words Action and Reaction as they occur in the enunciation of the third law of motion are explained to mean forces, that is to say, they are the opposite aspects of one and the same stress.

In the passage quoted above a new and different sense is given to these words by estimating Action and Reaction by the product of a force into the velocity of its point of application. According to this definition the Action of the external agent is the rate at which it does work. This is what is meant by the Power of a steam-engine or other prime mover. It is generally expressed by the estimated number of ideal horses which would be required to do the work at the same rate as the engine, and this is called the Horse-Power of the engine.

When we wish to express by a single word the rate at which work is done by an agent we shall call it the Power of the agent, defining the power as the work done in the unit of time.

The use of the term Energy, in a precise and scientific sense, to express the quantity of work which a material system can do, was introduced by Dr Young (*Lectures on Natural Philosophy*, 1807, Lecture VIII).

In their interpretation, which includes aspects treated by Mach, notably with regard to Euler and the Bernoullis, Todhunter & Pearson (1886, vol. 1, paras. 33–55) suggest that Jacopo Riccati, pupil of Euler, has priority for the use of conservation of energy in connection with elastic strain (some time before 1754, since that is the year of Riccati's death). His work was apparently not published until 1761 (*Opera Conte Jacopo Riccati*, 1, Lucca). Also, he is quoted as associating *forza viva* (*vis viva*) with energy.

Todhunter & Pearson also suggest that the concept of conservation of energy can be detected in Euler's *Methodus inveniendi lineas curvas maximi minimive proprietate gaudentes* (1744). This is said, moreover, to exhibit the theologico-metaphysical tendency of mathematical investigations in the seventeenth and eighteenth centuries, which probably led to the principles of least action and (in addition) to the principle of least constraint which was subsequently established on a sound basis. (Euler apparently referred to Daniel Bernoulli's letter of October 1742 as including the proposition that the elastic curve adopted by a bar is such that is *vis potentialis* (potential energy) is a minimum.) But it is to Poncelet that the introduction of conservation of energy into applied or practical mechanics appears to be due (1831). (Bresse refers to *General* Poncelet in his *Mécanique appliquée* of 1859.) In Britain, Robison's reference to the 'power of strain' in his treatise on practical mechanics (Brewster, 1822) seems trivial.

Lamé and Clapeyron

Perhaps the earliest major contribution to the use of energy in theory of structures and elasticity is due to Clapeyron (Lamé & Clapeyron, 1831). Apparently Clapeyron's original memoir on the subject (c. 1826) was never published, but eventually (1858), a memoir on the work of elastic forces in an elastic solid deformed by the action of external forces appeared. It was not until the publication of Lamé's *Leçons* (1852) on theory of elasticity, however, that Clapeyron's theorem (as Lamé called it) became known (except to a few, including Saint-Venant, who refers to its history on p. 871 of his annotated edition (1883) of Clebsch's celebrated book). Essentially, it is merely a statement that the work done by the external forces on an elastic solid is equal to the strain energy of the solid, due to those forces, that is, that energy is conserved, a general concept as yet unrecognised by Clapeyron. The highly significant feature is Clapeyron's use of actual displacements as virtual displacements, for the purpose of deriving the theorem by the principle of virtual work, a principle which implies conservation of energy! He obtained, in fact, an equation which specified that twice the work done by the external forces is equal to twice the strain energy associated with the stresses in the elastic solid, due to those forces. It was the use of actual elastic displacements as virtual displacements, in this manner, which provided the basis for much of Mohr's outstanding work on analysis of statically-indeterminate elastic frameworks (Chapter 5) and which has become perhaps the single most powerful device in structural analysis. Soon afterwards there was what

Love (1892, 1927) referred to as the 'revolution' effected by the mathematician Green (1839) on the basis of *vis viva* or what is now known as the 'principle of conservation of energy' which Crotti (1888) also acknowledges (Chapter 10). Green stated the following principle:

> In whatever way the elements of any material system may act upon each other, if all the internal forces exerted be multiplied by the elements of their respective directions, the total sum for any assigned portion of the mass will always be the exact differential of some function. But this function being known, we can immediately apply the general method given in the *Mécanique analytique*, and which appears to be more especially applicable to problems that relate to the motions of systems composed of an immense number of particles mutually acting upon each other. One of the advantages of this method, of great importance, is that we are necessarily led by the mere process of the calculation, and with little care on our part, to all the equations and conditions which are *requisite* and *sufficient* for the complete solution of any problem to which it may be applied.

It relates in fact to the potential energy, of a strained elastic body, expressed in terms of strains; the differential coefficients of those strains being the components of stress. Green believed the function to be capable of expression in terms of powers and products of components of strain. He used the principle to deduce the equations of elasticity of a solid body for the general case involving twenty-one constants.

W. Thomson (Lord Kelvin) argued (1857) the validity of Green's strain energy function on the basis of the first and second laws of thermodynamics, whereby Thomson deduced that, in the absence of temperature change, the components of stress of a solid body are the differential coefficients of a function of the components of strain with respect to these components severally.

It seems that Clapeyron (1858) applied his theorem for the purpose of showing that, for a framework whose bars (of the same material) have cross-sections which are proportional to maximum stress sustained, being either in direct tension or compression, and if a load P causes that maximum (safe) stress T in all bars and deflexion f:

$$Pf = EVT^2 \qquad [(7.1)]$$

where V is the total value of material in the framework and E is the modulus of elasticity. Lamé considered this kind of application of the theorem in some detail and Moseley (1843), ascribing the equation to Poncelet (1831), criticises it for a numerical error.

The seventh 'Leçon' of Lamé (1852) is devoted to elastic energy ('work of elastic forces') with particular reference to Clapeyron's theorem, its derivation and its uses. Thus, in the paragraph 'Application to structures', Lamé considers frameworks made of bars of wood or iron and, applying Clapeyron's theorem, writes:

$$(9) \qquad \Pi a = \frac{E}{\sigma} F^2 l + \frac{E'}{\sigma'} F'^2 l' + \frac{E''}{\sigma''} F''^2 l'' + \dots \qquad [(7.2)]$$

where Π is the load applied to the framework and a the deflexion in its line of action; E, E', E'', ..., the moduli of elasticity of the various bars; σ, σ', σ'', ..., the cross-sectional areas; and l, l', l'', ..., the lengths of those bars; F, F', F'', ..., the forces sustained by the bars due to the load Π. Thus, since linear elasticity is involved, Lamé's equation (9) of his seventh 'Leçon' is a statement of Clapeyron's discovery that, for such elasticity, twice the work done by the loading is twice the elastic energy which that loading engenders in the elastic solid or structure. The possible value of the theorem with regard to economy of material is first discussed then demonstrated by the example of a simple triangular bar framework.

Lamé considers a framework ABC made of two similar inclined wooden bars $\overline{AB}$ and $\overline{AC}$ and a horizontal bar $\overline{BC}$ connecting them. The three bars have the same modulus of elasticity E. A vertical load Π is applied at the apex A and the deflexion of the framework there in its line of action is a. Each of the inclined bars sustains an axial force F such that $\Pi = 2F \cos \alpha$ if α is half the angle between those bars at A, so that $F = \Pi/2 \cos \alpha$; the horizontal bar sustains a tension of $F \sin \alpha = \Pi \sin \alpha/2 \cos \alpha$. If the bars have the same cross-sectional area σ and the length of $\overline{BC}$ is L so that the length of $\overline{AB}$ and $\overline{AC}$ is $l = L/2 \sin \alpha$ then, by his equation (9):

$$\Pi a = \frac{E}{\sigma} \left[\frac{L}{\sin \alpha} \left(\frac{\Pi}{2 \cos \alpha} \right)^2 + L \left(\frac{\Pi \sin \alpha}{2 \cos \alpha} \right)^2 \right] \qquad [(7.3)]$$

whence

$$a = \frac{EL\Pi}{4\sigma} \left(\frac{1 + \sin^3 \alpha}{\sin \alpha \cos^2 \alpha} \right) \qquad [(7.4)]$$

Lamé then proceeds to consider the value of α for a to be a minimum for the specified load Π and concludes that the condition is that ABC is an equilateral triangle. But he then suggests that if the various bars are to carry the same intensity of load (stress), then their cross-sectional areas should be proportional to their axial forces; then, if the cross-section of the inclined bars is σ, that of the horizontal bar should be $\sin \alpha$ and equation (9) gives:

$$a = \frac{EL\Pi}{4\sigma} \left(\frac{1 + \sin^2 \alpha}{\sin \alpha \cos^2 \alpha} \right) \qquad [(7.5)]$$

Ascribing the limiting stress the value Ω, then $F/\sigma = \Pi/2\sigma \cos \alpha = \Omega$

whence $\Pi/\sigma = 2\Omega \cos \alpha$, giving an expression for the limiting value of a as being:

$$a = \frac{EL\Omega}{2}\left(\frac{1+\sin^2 \alpha}{\sin \alpha \cos \alpha}\right)$$ [(7.6)]

If now it is desired that α is such that a is a minimum, it is found, on equating $da/d\alpha$ to zero, that $\tan \alpha = 1/\sqrt{2}$; 'that is to say that the height of the triangle *ABC* becomes half the diagonal of the square whose side is of length L; also the limit of Π, or of $2\Omega \cos \alpha$ is $2\sqrt{\frac{2}{3}}\Omega$; that of a becomes $2E\Omega h$'.

Review and conclusions. It is easy to treat complex assemblies of bars of wood or iron in the same way... In all those various examples the best design of the structure may be determined by Clapeyron's theorem, being a form of the *principle of work of elastic forces*. I believe that there has never been such a close approach to the general solution of the famous problem of *solids of uniform strength*, which concerned Girard, and of which nature has provided such remarkable examples. We can make use of the principle of work of elastic forces for a variety of problems of equilibrium of elasticity. The principle is clearly related to the well-known principle of *vis viva* in rational mechanics. Moreover, the two principles are equally useful...the new concept is a transformation rather than an extension of the original. The procedure used by M. Clapeyron to establish his theorem supports this interpretation in a striking manner.

Adopting Navier's method, Clapeyron reproduced the general and unique equation of the internal equilibrium of elastic solids using the principle of virtual velocities; then he substituted in the equation provided by that principle, the actual elastic displacements (u, v, w) as possible virtual displacements instead of $(\delta u, \delta v, \delta w)$.

Later, Lamé remarks in conclusion:

The contents of this Leçon actually lead to reflection and observation that Clapeyron's theorem does not embody a new principle, it is essentially an extension, a transformation of the principle of *vis viva* or, in modern parlance, of the principle of work. This extension is, moreover, an additional source of illumination to a principle which is already rich in potentialities. Does it, however, introduce as a possible consequence the abandonment or neglect of basic methods of rational mechanics? That would imply abandoning the creative instrument in favour of that which is created, being without status of its own. It is typical of the contributions of Navier, Coriolis and Clapeyron: geometers before engineers, who had recourse to mathematical analysis, to the original methods of rational mechanics, in order to solve the problems which they studied and publicised and used in the course of their daily work.

Lamé thus apparently acknowledged that Clapeyron's theorem was essentially the law of conservation of energy as applied to the behaviour of elastic solids or structures. That law or principle seemed to lack explicit recognition when Clapeyron appealed to the principle of virtual velocities for the solution of the particular problem from which, according to Saint-Venant (in his translation (1883) of Clebsch's book of 1862), his so-called theorem was identified. Indeed, it seems as though the wider implication of the principle of virtual velocities as indicative of a natural law of conservation, embracing energy in all forms, was hardly recognised as late as the middle of the century. Saint-Venant, in describing (1883) the origin of Clapeyron's theorem, dismisses, as trivial, Lamé's application of it to structures, having regard to the availability of the convenient and general formal methods of Navier and Clebsch for the analysis of elastic structures, irrespective of their degree of statical indeterminacy. Indeed, he apparently believed that the theorem was of very limited value by itself, a conclusion which is entirely in accordance with modern knowledge.

Moseley

Some ten years after Clapeyron derived his energy theorem, Moseley's book appeared (1843), in which it is evident that he was deeply interested in exploiting energy concepts. Indeed, he had come close to energy theorems in his attempts to establish new statical *extremum* principles (1833a, b) in which he discussed Euler's earlier attempt (1774) with reference to the problem of statical indeterminancy (in the nature of a solid with supernumerary supports). Moseley's book, based on lectures to engineering and architecture students at King's College, London, contains generous acknowledgement of Poncelet with regard to the calculation of the internal work (energy) of an elastic beam for a specific deflexion and the equality of that energy with the work done by the load which caused it. He quotes Poncelet's celebrated book (1831). Moseley's articles 368 and 369 are devoted to calculation of the internal work (strain energy) of beams and the results are in the form:

$$U = \frac{P^2}{EI} \times K \tag{7.7}$$

using his notation, where U denotes work; P is pressure (load); and EI is flexural rigidity. K is a constant which depends upon the precise nature of the problem and is introduced here for the sake of brevity. (Moseley's notation has clearly been retained to this day.)

In article 370 he considers the deflexion of a simply supported beam with linear elasticity caused by a single 'deflecting pressure'. The procedure adopted seems curious in retrospect. Thus he notes that if D is the deflexion under pressure P:

$$P = \frac{dU}{dD} \qquad\qquad [(7.8)]$$

(a result for which Castigliano is commonly accorded priority); then he writes:

$$P = \frac{dU}{dD} = \frac{dU}{dP}\frac{dP}{dD} \qquad\qquad [(7.9)]$$

and noting $U = K\, P^2/EI$, puts $dU/dP = 2K\, P/EI$ to give:

$$P = 2K\frac{P}{EI}\frac{dP}{dD} \qquad\qquad [(7.10)]$$

or

$$\frac{dD}{dP} = \frac{2K}{EI} \qquad\qquad [(7.11)]$$

finally, he integrates to obtain the form:

$$D = 2K\frac{P}{EI} \qquad\qquad [(7.12)]$$

But the required result could have been obtained simply by direct application of conservation of energy (of which Moseley was clearly aware, in view of his comments on Poncelet's work) thus:

$$\tfrac{1}{2}PD = U = K\frac{P^2}{EI} \qquad\qquad [(7.13)]$$

whence

$$D = 2K\frac{P}{EI}$$

for Moseley's method, as presented, was not capable of generalisation.

Moseley's apparent leaning toward the use of energy derivatives is impressive, the more so when it is recalled that he proposed what he called the principle of 'least resistance' in (1833a). Reference to this principle is made by Cotterill, whose contribution to energy theory is also highly significant and seems due to his knowledge of Moseley's work.

Menabrea

Before Cotterill's research (1865) into the use of energy methods in structural analysis, a formidable contribution was made in Italy by General L. F. Menabrea. Thus, in 1858, he communicated what he described as a 'New principle on the distribution of forces in elastic systems' to the French Academy, which was published under the heading of 'Correspondence' in the *Comptes Rendus* (vol. 46, p. 1056). The following is the substance of a translation of Menabrea's memoir which has hitherto been generally dismissed, with the notable exceptions of Mohr (chapter 10) and Cotterill (1892, who, surprisingly, accorded all credit to Menabrea in retrospect). It provoked others, notably Castigliano, to pursue the same objective. It is believed, however, to merit careful attention within the scope of the history of theory of structures. Thus, Menabrea writes:

This is the announcement of a new principle which I have called the Principle of elasticity: Whenever an elastic system assumes a state of equilibrium under the influence of external forces, the work due to the effect of the tensions or compressions of bars which connect the various nodes of the system is a minimum. The differential equation which expresses this minimum would be known as the *equation of elasticity*: it would be useful for the determination of tensions.

I will give in what follows a concise demonstration of this principle. We will consider the most general case, and let *n* be the number of nodes of the system connected by *m* elastic bars. Each individual joint or node is in equilibrium under the influence of the external forces applied to it and the tensions of the bars which meet there. The number of equations of equilibrium for *n* joints will be $3n$: if *p* is the number of those equations relating the external forces, regardless of the tensions, the number of equations which must contain the tensions reduces to $3n-p$. Also, in the event of *m* being $> 3n-p$, the foregoing equations are insufficient for determining all the tensions.

It would be the same if the system had a certain number of fixed points. This indeterminacy signifies that there would be an infinite variety of values of tensions which, together with specified external forces, are appropriate to keep the system in equilibrium. The values of the actual tensions depend on the elasticities of the respective bars, and when they are specified, so also are the tensions.

Since, in the case we are considering, the tensions may vary without destroying equilibrium, it must be allowed that such variations could take place independently of the total work of the external forces; they are always associated with elongations or contractions of the relevant bars, which take place in each of them, with development of work. The changes of length of bars considered, must be assumed to be very small and unaccompanied by deflexion of the joints. But because, during the small internal displacement, equilibrium is maintained and the work of the external forces is zero, it follows that the summation of the increments of work of the tensions is also zero.

In order to express this conclusion, suppose *T* the tension of any bar; δl the element of variation of its length: the work consequential on variation of tension which causes that change of length would be $T\delta l$, and so for the complete system we have:

(1) $\qquad \Sigma\, T\delta l = 0$ $\qquad\qquad\qquad\qquad\qquad$ [(7.14)]

letting *l* be the total extension or contraction which occurs originally, due to the tension *T*, we have, regardless of sign:

(2) $\qquad T = \epsilon l$ $\qquad\qquad\qquad\qquad\qquad\qquad\qquad$ [(7.15)]

where ϵ is a coefficient which I will call the 'relative coefficient of elasticity', and which is a function of the modulus of elasticity, the cross-section and length of the bars.

The work developed to cause this variation of length l will be equal to $\frac{1}{2}\epsilon l^2$, and hence the total work of the system will be equal to $\frac{1}{2}\Sigma \epsilon l^2$.

But by virtue of equations (1) and (2) we have:

(3) $\Sigma T\delta l = \Sigma \epsilon l\delta l = \delta \frac{1}{2}\Sigma \epsilon l^2 = 0$ [(7.16)]

this being the proof of the principle stated, concerning which we can again bring other considerations. It is equally possible to express it in an alternative manner because we have:

(4) $\Sigma T\delta l = \Sigma \frac{1}{\epsilon} T\delta T = \delta \frac{1}{2}\Sigma \frac{1}{\epsilon} T^2$ [(7.17)]

Also, the sum of the squares of the tensions divided by their relative coefficients of elasticity is a minimum. It is easy to prove that equations (3) and (4) correspond to a minimum and not a maximum.

The equation

(5) $\Sigma \frac{1}{\epsilon} T\delta T = 0$ [(7.18)]

is that which I describe by the name 'equation of elasticity'. We will proceed using this terminology.

The n joints of the system provide, as already stated, $3n-p$ equations of equilibrium containing the tensions. For the infinitesimally small variations of the tensions which are compatible with equilibrium, we would be able to identify, by reference to various values of T, the $3n-p$ equations which provide a means of eliminating an equal number of variations, δT from the equation of elasticity (5). We would equate to zero the coefficients of various variations δT remaining in equation (5). These coefficients, being functions of the external forces and the tensions themselves, the new equations, together with those of equilibrium, would be equal in number to those of the tensions to be found.

In substance, Menabrea is identifying virtual displacements within a structure, such that points of application of loads remain undisplaced (and accordingly he is simply dealing with equilibrium), a valid procedure which he proceeds to misuse in his attempt to identify a minimum elastic work theorem. The nature of his error becomes apparent when one refers to Fränkel's later work (1882). Menabrea quotes Véne and Dorna as subscribers to the principle which he believes he has succeeded in proving. It is interesting that Menabrea finally returned to the subject some twenty-six years later following controversy involving Castigliano, Cerruti and Sabbia. Bertrand indicated a correct proof (1869): Donati reviewed Menabrea's work in detail (*c.* 1890).

Finally, it is appropriate to note the interest which was apparent (immediately prior to Menabrea's paper) in the problem of solids with more than three points of support. Thus, in 1850, a paper by Francesco Bertelli, which dealt with the problem, appeared in the *Memoirs of the Academy of Sciences of Bologna*. The paper was in two parts, the first dating from 1843 and the second from 1844. In addition to a history of attempts to solve the problem, which includes Euler's attempt of 1774, Bertelli also suggests (as a novelty) that the theory of elasticity is an essential ingredient of a solution, and he describes an experimental device for studying the problem. Then, in the *Memoirs of the Academy of Sciences of Turin*, 1857, a paper by Dorna appeared on the same subject and this did, in fact, make a direct appeal to the theory of elasticity. Dorna's attempt at precise analysis was marred by the assumptions (hypotheses) which he introduced. It represented, however, a fundamental departure from the type of metaphysical approach which had been used as late as 1852 by G. Fagnoli and published in the *Memoirs of the Academy of Sciences of Bologna* (Todhunter, 1893).

Notes

Appreciation of the true nature of the law of conservation of energy seemed to take place after 1840 with the abandonment of the caloric theory and the reappraisal of Carnot's theory which culminated in the monumental work of Clausius on the mechanical theory of heat (*c.* 1850). Tait (1868) gives a useful historical survey which includes what has been called the 'Mayer–Joule controversy', with regard to the equivalence of mechanical work and heat. Apparently Mayer deduced it theoretically (*c.* 1842), while Joule determined it by experiment some six years later. Tait is critical of Mayer's claim but quotes the remarks of Helmoltz in support of it. Clapeyron's interest in Carnot's theory of heat and work (*c.* 1834) is recalled, and his reputation in the field of thermodynamics was such that after 1844 he became professor of steam engines at l'Ecole des Ponts et Chaussées. Then, some forty years later, Weyrauch published an appreciation of Mayer's work.

According to Todhunter (1886), the subject of virtual velocities was chosen by the Turin Academy of Sciences for a prize essay (J. F. Servois was the successful candidate).

According to Saint-Venant (1883, footnote, p. 871) Clapeyron derived his so-called theorem relating to elastic energy 'toward 1840 on the occasion of research in dynamics relating to the dissipation of the energy of wagon springs'. Saint-Venant remarks that Navier obtained essentially the same result in 1821 and that it is nothing more than an application of conservation of energy.

Menabrea finally attempted to justify his principle by seeking to show that it is in accordance with Levy's method (1874) as well as with *the principle of least work* which was deduced, he said, by Euler and which applied to the determination of the elastic curve. This he did in a note (1884) to the French Academy (whose Perpetual Secretary, Bertrand, seemed sympathetic to him).

Menabrea published a note on Babbage's calculating machine in *Comptes Rendus* (1884, vol. 99, pp. 179–81) and quoted a letter written by Babbage on 23 August 1843, which mentions Lady Lovelace, the only daughter of Byron.

8

The later development and use of energy principles

Exploitation of the doctrine of energy, using energy functions and their derivatives in theory of structures, seems to have begun in earnest on the continent of Europe (by coincidence) soon after Cotterill's three important articles appeared in 1865. It was primarily due to the researches and principles of the Italian railway engineer Castigliano (1873, 1879), after Menabrea (1858). The implicit objective was to remedy deficiencies of statics by means of conditions of compatibility of elastic strain. Indeed, Castigliano's so-called principle of least work (terminology of Menabrea, 1884) was to become perhaps the best-known general method of structural analysis toward the end of the century. The contributions of Fränkel, Crotti and Engesser to the energy approach are, however, significant. But, in restrospect, Cotterill's priority over Castigliano and others seems unquestionable after careful study of his original articles. His obscurity, until comparatively recently, is undoubtedly due to the fashion in Britain to publish original work in both pure and applied science in journals devoted to natural philosophy, outstanding among which is the *Philosophical Magazine* in which Moseley and Maxwell, as well as Cotterill, published their contributions to theory of structures. Originally, like Moseley, a Cambridge mathematician of St John's College, Cotterill (who lived from 1836 until 1922) became professor of applied mathematics at the Royal Naval College.

Cotterill, 1865

Cotterill appeared to seek to rationalise and utilise the concept of natural economy which was apparently manifest in the condition for minimum strain energy of elastic systems; while also recognising that the approach he adopted identified those conditions with compatibility of strains (as distinct from conditions of equilibrium). His various applications included those to arch ribs and thick cylinders (where he noted that he obtained results similar to those of Lamé and Rankine), and he used Lagrange's method for minimising integrals. He noted, moreover, that Daniel Bernoulli (Chapter 7) had prescribed minimum elastic energy, in the same form, with regard to the behaviour in bending of a thin elastic strip (seemingly on a purely theologico-metaphysical basis, that is, an unqualified belief in natural economy). The essence of Cotterill's contribution is to be found in the first two of his three articles.

He published, first, an article 'On an extension of the dynamical principle of least action' (1865*a*) and the essential features of his proposition are contained in the following quotation which refers to Moseley's early article (1833*a*):

Now Mr. Moseley has shown that if any number of pressures are in equilibrium, some of which are resistances, then each of these resistances is a minimum, subject to the conditions imposed by the equilibrium of the whole – a principle which he has called the principle of Least Resistance; let us assume this principle, and let us further suppose, for the present, that it is generalised so as to include the case of resisting forces generated, as above described (by elastic displacements); then each of those resisting forces is a minimum, subject to the general conditions stated above; and, further, the relative displacements which are the cause of those forces must also be the least possible, and the work done the least possible. Thus in the assumption mentioned, to which I shall return in the sequel, it appears that the work done is a minimum, subject to the law of conservation of energy and the statical conditions of equilibrium; and this principle, analogous to the dynamical principle of Least Action, it is the object of this article to consider and apply.

If the work done be expressed in terms of the resisting force at all points of the system, or some of them, then, the law of conservation of energy being implicitly satisfied, we have simply to make the work done a minimum, subject to the statical conditions of equilibrium.

Cotterill demonstrates his proposition by first considering a uniform beam loaded by a distributed load of uniform intensity w over its length and subjected to restraining couples M_1 and M_2 at its ends. If $2c$ is the span, the strain energy U of the loaded beam is given as:

$$U = \frac{c}{3EI}\left\{M_1^2 + M_1 M_2 + M_2^2 - wc^2(M_1 + M_2) + \tfrac{2}{5}w^2c^4\right\} \qquad [(8.1)]$$

Considering the problem of determining the values of M_1 and M_2 which would cause the ends of the beam to be fixed horizontally, Cotterill proceeds by asserting that the values of M_1 and M_2, which cannot be found by statics alone, must be such that U is a minimum, that is:

$$2M_1 + M_2 - wc^2 = 0; \quad 2M_2 + M_1 - wc^2 = 0 \qquad [(8.2)]$$

whence

$$M_1 = M_2 = \tfrac{1}{3}wc^2 \qquad\qquad [(8.3)]$$

which result, he remarks,

agrees exactly with that given by the ordinary method...the reason of which will be seen from what follows. Let i_1, i_2 be the slopes at the extremities of a beam acted on by M_1, M_2 at its extremities, and by the uniform load w; then U being, as shown above, an homogeneous quadratic function of M_1, M_2, we have:

$$2U = \frac{\mathrm{d}U}{\mathrm{d}M_1}M_1 + \frac{\mathrm{d}U}{\mathrm{d}M_2}M_2 + \frac{\mathrm{d}U}{\mathrm{d}w}w \qquad\qquad [(8.4)]$$

but by the law of conservation of energy:

$$2U = M_1 i_1 + M_2 i_2 + wu \qquad\qquad [(8.5)]$$

Where u is the 'area of deflexion' of the beam; and, comparing these expressions, we see that:

$$\frac{\mathrm{d}U}{\mathrm{d}M_1} = i_1; \quad \frac{\mathrm{d}U}{\mathrm{d}M_2} = i_2; \quad \frac{\mathrm{d}U}{\mathrm{d}w} = u \qquad\qquad [(8.6)]$$

but the ordinary method is founded on the consideration that the beam is horizontal at its extremities, in other words, that $i_1 = 0$, $i_2 = 0$; so that the two methods lead to the same result by the same equations. And this will be the case in all questions concerning continuous beams; but the present method enables us to obtain the requisite equations by differentiation of a single function.

Toward the end of his article, Cotterill proceeds as follows:

Having briefly indicated the mode of applying the principle of Least Action to various problems, I return to its demonstration.

From the description given of the process by which equilibrium is attained, it is apparent that if the principle of Least Resistance be given, the principle of Least Action follows, and vice versa; and since the principle of Least Resistance is well known, I have assumed it. But inasmuch as that principle has perhaps never been satisfactorily proved, at least in its general form, it will be well to give a direct demonstration of the principle of Least Action in the case where the body is perfectly elastic.

Let X, Y, Z be the components of one of the forces acting on a free perfectly elastic body; u, v, w the displacements of its point of application parallel to three rectangular axes; U the work done in the body, then:

$$2U = \Sigma(Xu + Yv + Zw) \qquad\qquad [(8.7)]$$

but U may be expressed as a homogeneous quadratic function of the forces; therefore:

$$2U = \Sigma\left\{X\frac{\mathrm{d}U}{\mathrm{d}X} + Y\frac{\mathrm{d}U}{\mathrm{d}Y} + Z\frac{\mathrm{d}U}{\mathrm{d}Z}\right\} \qquad\qquad [(8.8)]$$

comparing expressions U, we see that:

$$\frac{\mathrm{d}U}{\mathrm{d}X} = u; \quad \frac{\mathrm{d}U}{\mathrm{d}Y} = v; \quad \frac{\mathrm{d}U}{\mathrm{d}Z} = w \qquad\qquad [(8.9)]$$

Now conceive the body, instead of being free, to be immovably attached at certain points to some fixed object, then we shall have for these points:

$$\frac{dU}{dX} = 0; \quad \frac{dU}{dY} = 0; \quad \frac{dU}{dZ} = 0 \qquad [(8.10)]$$

that is, the variation in U, due to a change in the resisting force at the fixed boundaries of the system, is zero. And the same is true for a change in the resisting force anywhere within the mass; for conceive the body to be divided into two parts by a surface of any form passing through the point, and let U_1, U_2 be the works done in the two portions, then $\frac{dU_1}{dX}$ and $\frac{dU_2}{dX}$ are evidently equal and of opposite sign, that is, $\frac{dU}{dX} = 0$ as before. Since, then, the change in U, consequent on any possible change in the resisting forces, is zero, U must be a minimum (the other two possible hypotheses being easily seen to be inadmissible), and the principle is proved for a perfectly elastic body or system of bodies.

Cotterill appears to recognise that his minimisation of U provides conditions of compatibility of deformations.

Being apparently dissatisfied with his justification of the energy method, Cotterill returns to the matter at the end of his second article (1865*b*) 'On the equilibrium of arched ribs of uniform section', where he writes:

In my former article I endeavoured to show that if X, Y, Z be components of one of the forces acting on a perfectly elastic body; u, v, w the displacements of its point of application produced by the action of the forces on the body, then:

$$\left. \begin{array}{l} dU/dX = u \\ dU/dY = v \\ dU/dZ = w \end{array} \right\} \qquad [(8.11)]$$

But the reasoning is not so conclusive as the following. Since

$$2U = \Sigma (Xu + Yv + Zw) \qquad [(8.12)]$$

then

$$2\delta U = \Sigma (X\delta u + Y\delta v + Z\delta w) + \Sigma (u\delta X + v\delta Y + w\delta Z) \qquad [(8.13)]$$

but $\Sigma (X\delta u + Y\delta v + Z\delta w)$ is the increment of the energy expended, by which the law of conservation of energy is equal to δU the increment of work done, therefore, we have also:

$$\delta U = \Sigma (u\delta X + v\delta Y + w\delta Z) \qquad [(8.14)]$$

whence the above equations (8.11) follow.

He also notes that $dU/du = X$; $dU/dv = Y$; $dU/dw = Z$ (the energy derivative obtained by Moseley in his book, 1843). Equation (8.14) is the basis of the true justification of the use of strain energy to derive deformations, being actually the expression for the variation of complementary energy which was introduced explicitly by Engesser (1889). There is some essential degree of similarity of Cotterill's derivations with those of Castigliano, including the principle in question (so-called least work) as is evident in the following paragraphs.

Castigliano, 1873

In the preface to his book (1879), Castigliano writes as follows:

In the year 1818 Captain Véne of the French Engineer Corps enunciated a principle which was absolutely incorrect under the conditions to which he wished to apply it, but which, by one of those peculiar combinations of circumstances of which science presents several examples, was destined to lead later to the discovery of the theorem of least work.

After this first step, the distinguished scientists Messieurs A. Cournot, Pagani, Mossotti, A. Dorna, and General L. F. Menabrea investigated the question. The last-mentioned gave the name 'principle of elasticity' to the theorem of least work, and made it the subject of his researches, in a first memoir presented in 1857 to the Academy of Science of Turin, later in a second presented in 1858, to the Academy of Science of Paris, and again in a third submitted in 1868 in the Turin Academy. Since, however, the proofs given by M. Menabrea were not exact, the 'principle of elasticity' was not accepted by the greater number of the authorities, and some of them published memoranda to show the fallacy of it. It was not until 1873 that we gave, in our above-mentioned thesis, the first rigorous proof, in a form which appeared to us clear and exact, of the theorem of least work.

Castigliano nobly conceals his dispute (in published correspondence) with Menabrea, which involved the then President of the Turin Academy, Professor Cremona.

This book (1879) is a substantial work and includes aspects of theory of elasticity as well as theory of bar frameworks and arches. (It is, incidentally, evident that Castigliano knew of Navier's method for statically-indeterminate bar frameworks.) In chapter 1 of Andrews' translation (1919) the work equation (15) is:

$$W_i = \tfrac{1}{2}\Sigma F_p r_p \qquad\qquad [(8.15)]$$

where W_i is described as the internal work of a linearly elastic structure and F_p and r_p respectively, are applied load and associated deflexion in its line of action. He had, in fact, obtained an expression for the internal work of such a structure, as a function of the external forces, having calculated the corresponding expression in terms of forces in the members (of a pin-jointed framework, in fact). Castigliano's procedure here seems unduly complex and rigorous in the absence of simple application of the law of conservation of energy (which he quotes elsewhere).

His progress toward the 'theorem of least work' embraces his theorem of the differential coefficients of the internal work, part 1 of which is:

$$F_{\text{p}} = \frac{\mathrm{d}W_i}{\mathrm{d}r_{\text{p}}} \qquad\qquad\qquad [(8.16)]$$

(already noted by Moseley) while part 2 is:

$$r_{\text{p}} = \frac{\mathrm{d}W_i}{\mathrm{d}F_{\text{p}}} \qquad\qquad\qquad [(8.17)]$$

Proof of the former is trivial but to obtain the latter he considers the differential of equation (8.15):

$$\mathrm{d}W_i = \tfrac{1}{2}\Sigma F_{\text{p}}\mathrm{d}r_{\text{p}} + \tfrac{1}{2}\Sigma r_{\text{p}}\,\mathrm{d}F_{\text{p}} \qquad\qquad\qquad [(8.18)]$$

or

$$\Sigma F_{\text{p}}\,\mathrm{d}r_{\text{p}} = \tfrac{1}{2}\Sigma F_{\text{p}}\,\mathrm{d}r_{\text{p}} + \tfrac{1}{2}\Sigma r_{\text{p}}\,\mathrm{d}F_{\text{p}} \qquad\qquad\qquad [(8.19)]$$

whence, he concludes

$$\Sigma F_{\text{p}}\,\mathrm{d}r_{\text{p}} = \Sigma r_{\text{p}}\,\mathrm{d}F_{\text{p}} \qquad\qquad\qquad [(8.20)]$$

and since

$$\mathrm{d}W_i = \Sigma \frac{\mathrm{d}W_i}{\mathrm{d}F_{\text{p}}}\mathrm{d}F_{\text{p}} = \Sigma r_{\text{p}}\,\mathrm{d}F_{\text{p}} \qquad\qquad\qquad [(8.21)]$$

then

$$r_{\text{p}} = \frac{\mathrm{d}W_i}{\mathrm{d}F_{\text{p}}} \qquad\qquad\qquad [(8.22)]$$

After some further attention to the theorem of the differential coefficients of internal work, Castigliano addresses himself to the theorem of least work. First, he specifies that W_i is the internal work of a frame which is reduced to contain only its essential bars and that it is a function of the external forces and of the forces in the omitted (redundant) bars. Then if N_p and N_q are two joints connected by one of the latter and T_{pq} is the tension in it, it follows that the differential coefficient of W_i with regard to T_{pq} expresses the amount by which the nodes approach each other. Again, that bar extends by an amount T_{pq}/ϵ_{pq} (where $\epsilon_{pq} = A_{pq}E_{pq}/L_{pq}$) so that:

$$\frac{\mathrm{d}W_i}{\mathrm{d}T_{pq}} = -\frac{T_{pq}}{\epsilon_{pq}} \qquad\qquad\qquad [(8.23)]$$

$$\frac{\mathrm{d}W_i}{\mathrm{d}T_{pq}} + \frac{T_{pq}}{\epsilon_{pq}} = 0 \qquad\qquad\qquad [(8.24)]$$

Noting there will be a similar equation for each of the omitted bars, sufficient equations will be obtained in the unknowns T_{pq}. Since the internal work of the original structure (with all bars present) is given by the formula:

$$W_i + \frac{1}{2}\Sigma\left(\frac{T^2_{pq}}{\epsilon_{pq}}\right) \qquad\qquad\qquad [(8.25)]$$

the equations derived, which express the geometrical conditions which the strains must satisfy, are the equations to zero of the differential coefficients of the internal work for the whole structure: therefore the stresses which occur after strain are those which make this work a minimum.

There is little doubt that Castigliano's least work principle is among the best-known aspects of theory of structures. He extended the method to take account of self-straining, including thermal effects, and showed that if such an effect is equivalent to an initial lack of fit (shortness) of a redundant bar in line pq of λ_{pq}, then for the whole structure the function:

$$W_i + \frac{1}{2} \Sigma \left(\frac{T^2_{pq}}{\epsilon_{pq}} \right) - \Sigma T_{pq} \lambda_{pq} \qquad [(8.26)]$$

is a minimum.

It is clear, however, that Cotterill anticipated Castigliano (by eight years) with regard to the differential coefficients of the internal work (strain energy) of a structure. Also, there is a certain similarity in the approaches which each used. Their treatment of the minimum strain energy theorem is different, however, although the nature of the equations obtained by means of the theorem (equations of compatibility of strain) seemed clear to both. Linearity of elasticity of structures and their bars was assumed by Cotterill and Castigliano, thus implying constancy of the geometry of structures (that is, small deformations) and materials of construction which obey Hooke's law. It is, incidentally, noteworthy that, prior to energy concepts in his book, Castigliano discoursed on the explicit use of conditions of equilibrium and compatibility of strains, together with the law of elasticity, for dealing with statically-indeterminate frameworks.

Crotti, 1888

Francesco Crotti was apparently Castigliano's friend and colleague. His book on theory of elasticity, including fundamental principles and application to structures, appeared in 1888 (although in the book he refers to lectures given at Milan in 1883). It is mathematically sophisticated by comparison with the writings of Cotterill, Castigliano or Engesser. Indeed, having specified perfect (linear) elasticity, results of theorems (including the reciprocal theorem and elastic coefficients) are derived by mathematical techniques and devices: the relevant terminology and concepts are in the language of the mathematician rather than that of the natural philosopher. But there are applications of the theorems to practical problems and at the beginning of the book there is a useful survey of the historical development of theory of elasticity, together with an account which is similar to Love's account (1892) of Green's contribution to the doctrine of energy. Crotti also gives much credit to Navier generally. The mathematical rigour and use of variational techniques is impressive: the duality principle (ascribed to Helmoltz) is noted, and theory of stability of equilibrium is considered. Attention here is confined to only the essential features of Crotti's treatment of energy concepts: other aspects of his work are considered in Chapter 10.

Crotti gives alternative expressions for the work (L) by forces applied to an elastic structure:

$$L = \phi(f_1, f_2, \ldots, f_n) \qquad [(8.27)]$$

and

$$L = \psi(F_1, F_2, \ldots, F_n)$$

where F is an individual force and f is the elastic deflexion in the line of action of that force. Subsequently, Crotti states that the first differential of work is:

$$F_1 \, df_1 + F_2 \, df_2 + \ldots \qquad [(8.28)]$$

whence

$$\frac{dL}{df_1} = F_1; \quad \frac{dL}{df_2} = F_2, \ldots \qquad [(8.29)]$$

Then he proposes a quantity:

$$\lambda = F_1 f_1 + F_2 f_2 + \ldots - L \qquad [(8.30)]$$

whence

$$d\lambda = f_1 \, dF_1 + f_2 \, dF_2 + \ldots + (F_1 \, df_1 + F_2 \, df_2 + \ldots - dL) \qquad [(8.31)]$$

or since, by conservation of energy, the quantity in parentheses is zero:

$$d\lambda = f_1 \, dF_1 + f_2 \, dF_2 + \ldots \qquad [(8.32)]$$

which gives

$$\frac{d\lambda}{dF_1} = f_1, \quad \frac{d\lambda}{dF_2} = f_2, \ldots \qquad [(8.33)]$$

Crotti subsequently notes that the assumption of linear elasticity and behaviour implies that:

$$2L = \Sigma fF \qquad [(8.34)]$$

and since, by definition, $L + \lambda = \Sigma fF$, then $\lambda = L$. He proceeds to discuss the implications and states without further proof, that λ may be substituted for L in Castigliano's theorem of least work and suggests that the theorem is then applicable when elasticity is non-linear.

Engesser, 1889

Engesser's article on the proposed method of complementary energy was published in a journal addressed to engineers. He had become professor at the Karlsruhe Polytechnikum in 1885 when he was thirty-seven years of age and became distinguished for his work on non-linear phenomena, including buckling. He wrote:

The following considerations refer to the behaviour of statically-indeterminate frames with optional deflexion laws. The theorem of virtual work is the safest and most convenient way of solving frame problems for which the theorem of least work, whose validity is restricted to the case of specified deflexion laws, is inadequate. The universal theorem of minimum complementary energy will be introduced instead of this.

The essence of Engesser's derivation is given below, using his notation and giving his equation numbering on the left.

If the deflexions are assumed to be small so that the geometry of the deflected structure does not differ materially from the original, the principle of virtual work gives:

(1) $\Sigma Pv = \Sigma Se + \Sigma Cc$ [(8.35)]

$$\text{Here: load} \quad = P \; \left.\begin{array}{l}\\ \\ \\\end{array}\right\}$$

Here: load $= P$ ⎫
reaction $= C$ ⎬ and corresponding deflexions $\begin{array}{l} v \\ c \\ e \end{array}$
bar force $= S$ ⎭

The bar forces and reactions in a statically-indeterminate structure with m redundants, represented by $X', X'', \ldots, X^m$, are given by:

(2) $S = S' + s' X' + s'' X'' + \ldots + s^m X^m$ [(8.36)]
$C = C' + c' X' + c'' X'' + \ldots + c^m X^m$

where

S' is the force in a bar when redundants are removed, produced by P

s' is the force in a bar when redundants are removed, produced by $X' = 1$

s^m is the force in a bar when redundants are removed, produced by $X^m = 1$

and

C, c', c^m are corresponding quantities for a reaction or support.

After considering a variety of analytical details Engesser considers a general relationship between strain ϵ and stress σ of a bar, whereby $\epsilon = f(\sigma)$, so that when a bar of length s and cross-sectional area F is subjected to a force S:

$e = s\epsilon = sf(\sigma) = sf(S/F)$ [(8.37)]

and if the temperature of the bar is simultaneously raised by t degrees:

$$e = (\epsilon+\alpha t)s = \{f(S/F)+\alpha t\}s \qquad\qquad [(8.38)]$$

Also, the effect of lack of fit λ of a bar may be equivalent to a small extension or rise of temperature $t = \lambda/\alpha s$.

Having given further analytical details regarding 'self-straining', Engesser makes the important statements:

The theorem of minimum strain energy derived by Castigliano and Fränkel, on the assumption that Hooke's law, $\epsilon = \sigma/E$, is obeyed, is not valid for the optional deflexion law $\epsilon = f(\sigma)$.

As the bar force increases steadily from 0 to S [Fig. 52] the work $a = \int_0^e Sde$ and so the work of all the bars is:

$$A = \Sigma a = \Sigma \int_0^e Sde \qquad\qquad [(8.39)]$$

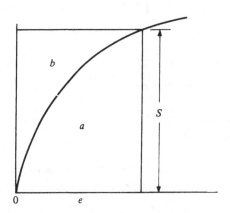

Fig. 52. From Engesser (1889).

If the bar force S were to traverse the distance with constant magnitude, then the work performed, which may be called the virtual work is $a_v = Se$ and the total virtual work $A_v = \Sigma a_v = \Sigma Se$.

The difference between the virtual work A_v and the real work A can be called the 'complementary work' designated by the symbol B. We then have:

$$B = A_v - A = \Sigma a_v - \Sigma a = \Sigma b = \Sigma \int_0^S edS \qquad\qquad [(8.40)]$$

It is also noted (by Engesser) that, since $A_v = \Sigma Pv$, then $\delta A_v = \Sigma P\delta v + \Sigma v\delta P = \delta A + \delta B$, and that $dB/dP = v$, the deflexion of load P; while with regard to a force X, due to a redundant bar, $dB/dX = 0$, so that Engesser is able to write:

The differential coefficients of B with regard to X, in turn, are set equal to zero, giving equations of the form:

(10) $$0 = \frac{\partial B}{\partial X} = \Sigma \frac{\partial}{\partial X} \int dS\, e = \Sigma \frac{\partial S}{\partial X}\, e$$ [(8.41)]

Equation (8.41) 'is the condition for the minimum value of B as a function of the quantities X. The resulting values of X correspond with the minimum of complementary energy B'.

The complementary energy b for a single bar subjected to a load and

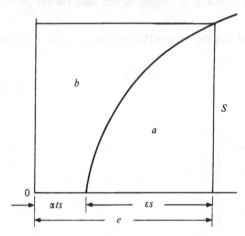

Fig. 53. From Engesser (1889).

temperature change is shown in Fig. 53. If the work curve is a straight line, following Hooke's law, then with an arbitrary temperature rise, t:

$$B = \Sigma b = \Sigma S\left(\frac{es}{2} + \alpha ts\right) = \Sigma S\left(\frac{Ss}{2EF} + \alpha ts\right)$$ [(8.42)]

which is the same as what Müller-Breslau (1886b, *Die neueren Methoden*, pp. 185ff) called the ideal strain energy A_1. The theorem of the minimum value of the ideal strain energy is therefore a special case of the theorem of minimum complementary energy.

If the strain curve is a parabola $\epsilon = (1/C)\sigma^n$, then with $t = 0$

$$\left.\begin{array}{l} B = \Sigma b = \Sigma \dfrac{1}{n+1} Ss \\[2ex] A = \Sigma a = \Sigma \dfrac{n}{n+1} S\epsilon s = nB \end{array}\right\}$$ [(8.43)]

If A and B reach their minimum values simultaneously, then the condition min A can be used in place of min B as in the earlier derivation.

Elegant, illustrative examples of the use of the minimum energy principle (in the manner of Castigliano, as well as of Crotti and Engesser, if elasticity is linear throughout) are given in the celebrated article by Mohr's colleague at Dresden, Fränkel (1882), to which reference is made in Chapter 3, with regard to his analysis of arches and suspension bridges. These examples are described in Chapter 9.

Fränkel, Müller-Breslau, Weyrauch

Fränkel derived the minimum energy principle of 'least work' independently of Castigliano (as noted by Engesser, 1889), for in his article he acknowledges Winkler's advice with regard to Castigliano's priority (received only after his manuscript had gone to the printer!). It seems, moreover, that Fränkel's article (1882) made Müller-Breslau aware of Castigliano's principle. Fränkel's derivation is significantly different from that of Castigliano: indeed, it bears some conceptual resemblance to Menabrea's derivation. Having remarked on the evidence for *extremum* principles, for example, Winkler's principle for arches, he considers small changes in the forces in the bars of a structure, which satisfy the conditions of equilibrium if there are no changes in the loads or forces applied to the structure and, in particular, the variation of work (virtual work) if compatible displacements, identical to those caused by the loads, are introduced at the same time. Since there are no changes in the loads, that equation specifies that the sum of the actual elastic changes in length of the bars, multiplied by the small changes in bar forces, is zero. Expressing those elastic changes in length in terms of the bar forces, he examines the number of additional equations which may then be derived to supplement the conditions of equilibrium when the framework is statically-indeterminate and concludes that it is sufficient to enable analysis of the structure to be achieved (though without referring specifically to the physical implication of compatibility of strain). It is now known that Bertrand suggested to Menabrea (in a letter, 1869) similar details for the purpose of validating his principle but, apparently, this was unknown to Fränkel.

A few years after Fränkel's paper appeared, Weyrauch's treatment of the doctrine of energy, in his book (1884) on theory of elasticity, included reference to the work of Clapeyron, Castigliano and Fränkel and, moreover, implicitly identified the quantity, complementary energy, but did not exploit its analytical value. Thus, he considers the virtual work represented by the sum of products of impressed forces and the relevant elastic displacements of a solid, denoting it by D. Then he considers the total variation of that quantity and concludes that:

$$\delta D = \delta_v D + \delta_s D \qquad [(8.44)]$$

where $\delta_v D$ is an element of what is now known as complementary energy, and $\delta_s D$ is an element of strain energy. Weyrauch's treatment of energy is of substantial general interest because it includes thermodynamic aspects, on the one hand, and application of Castigliano's principle to framework analysis in comparison with Mohr's method, on the other.

Weyrauch, with Fränkel, Müller-Breslau and others, participated in a protracted controversy with Mohr about the value of Castigliano's principle. Apparently this was one aspect of the differences between Mohr and Müller-Breslau, which is discussed in Chapter 10. (Mohr plainly accorded priority to Menabrea in an article published in 1883 and mentioned Castigliano only some time later.)

All of the energy principles considered herein have the common implied objective of deriving deflexions and, for statically-indeterminate systems, conditions for compatibility of strains. Those strains may be due to thermal effects in addition to applied loads. They are, in fact, alternative methods of obtaining precisely the same kind of equations as those which Maxwell and Mohr derived (with the aid of 'Clapeyron's theorem' and virtual work respectively) and, moreover, they avoid the intellectual effort required for Levy's method. But they are concerned with a different approach from that which Navier adopted for statically-indeterminate systems of bars. It is strange to note that the energy theorem relevant to his approach already existed (as noted above) in classical mechanics, namely the principle of minimum potential energy for equilibrium. It seemed to be overlooked with regard to theory of structures, for a variety of reasons including, perhaps, those for which Navier's method remained obscure during the nineteenth century. Thus the nature of the structural forms then in use was such that the degrees of freedom of deflexion of joints were far more numerous than the supernumerary (redundant) elements (bars or joints), so that the method might seem invalid. That principle was to emerge in the twentieth century, largely as a means of obtaining rapid approximate solutions to certain types of complex continuous structures. But careful study of Crotti's work (1888) indicates that he was aware of the principle of minimum potential energy and, indeed, discusses it in relation to the problem of stability of equilibrium of systems. Having regard to his explicit recognition of the duality aspect, including, for example, the use on the one hand of flexibility coefficients and, on the other, of stiffness coefficients, with regard to the behaviour of linearly elastic structures, it is tempting to conclude that he simply saw no advantage in applying the principle of minimum potential energy to structural analysis.

Notes

It is surprising that Cotterill seemed to lack confidence in his contribution to energy principles (1865) to the extent that, when he eventually became aware of Menabrea's contribution, he accorded (1892) all credit to him.

The introductory remarks of Fränkel's article are interesting: he comments on the practice of avoiding (in cases where statics proves insufficient) the use of theory of elasticity by attempting to replace it with an appeal to principles 'more or less theoretical', which depend upon some principle of minimum force.

Fränkel (1882) actually used the concept of elastic energy and work rather than the principle of virtual work.

9

Applications of the least work principle: elastic theory of suspension bridges

This chapter deals with the elegant and especially significant applications of the least work principle described by Fränkel (1882). These applications almost certainly originate from his friendship and collaboration with Winkler. Thus, he begins with the relatively difficult problems of the elastic arch and suspension bridge. In so doing, it seems that he was mindful of Winkler's principle (Chapter 3) for the thrust-line of an arch (1879a). Indeed, that principle probably led to his search for a means of establishing a principle of least work for elastic structures generally, the successful outcome of which was marred by the discovery, just as his work was poised for publication, that he had been anticipated by Castigliano. Thus he acknowledges Winkler for that information and it may be judged by the fact that Winkler himself had then only recently become aware of Castigliano's work through the French (1880) edition of the original work.

It will be noted that, throughout application of the least work principle to an arch, Fränkel contrasted the results with those derived by Winkler. Having thus discussed the principle in relation to what he termed 'the elegant work' of Winkler, he turned his attention to what he identified as the closely related problem of the stiffened suspension bridge (incorporating inversion of the arch) and succeeded in determining the elastic theory in advance of Levy's celebrated treatment (Chapter 3). Details of these two applications are given herein as being little known and of substantial historical interest in addition to being generally instructive.

Elastic arch

Having derived the formula for the strain energy A of an elastic arch of flexural rigidity EJ due to a bending moment M, as:

$$A = \frac{1}{2} \int \frac{M^2 ds}{EJ} \qquad [(9.1)]$$

where ds is an element of length, Fränkel considers Winkler's problem of an arch of shape defined by ordinates y, subjected to a uniformly distributed load of intensity q per unit horizontal length of span. The line of thrust is defined by ordinates η while the horizontal thrust at any section is designated by H. Having specified that $d^2y/dx^2 = q/H$ for the arch, and noting that η is the variable in respect of y, Fränkel suggests that by integration of that equation:

$$\eta = B + Cx + \frac{f(x)}{H} \qquad [(9.2)]$$

where B and C are constants of integration. It is specified next that the bending moment at a point (x, y) is:

$$M = H(\eta - y) = H\left(B + Cx + \frac{f(x)}{H} - y\right) \qquad [(9.3)]$$

and that for a uniform arch the strain energy is represented by:

$$S = \int M^2 ds = \int H^2(\eta - y)^2 ds$$
$$= \int (BH + CHx + f(x) - Hy)^2 ds = \text{minimum} \qquad [(9.4)]$$

Then, in parentheses, it is remarked that Winkler's principle specifies that $\int (\eta - y)^2 ds = \text{minimum}$. The necessary conditions for equation (9.4) to be a minimum are given as:

$$\frac{dS}{dB} = 2 \int H^2(\eta - y) ds = 0 \quad \text{or} \quad \int (\eta - y) ds = 0$$

$$\frac{dS}{dC} = 2 \int H^2(\eta - y) x ds = 0 \quad \text{or} \quad \int (\eta - y) x ds = 0 \qquad [(9.5)]$$

$$\frac{dS}{dH} = 2 \int H(\eta - y)(B + Cx - y) ds \quad \text{or} \quad \int (\eta - y) y ds = 0$$

having regard to equations (9.3) and (9.4).

Also, if the arch is not uniform then the least work condition is that:

$$\int \frac{M^2 ds}{J} = \text{minimum}$$

or, according to Winkler's principle

$$\int \frac{(\eta - y)^2}{J^2} ds = \text{minimum} \qquad [(9.6)]$$

It is to be noted here that Straub (1952) incorrectly gives the date of Winkler's principle as 1867 and that Timoshenko's (1953) account of it is different from the original version.

The principle of least work is thus taken to indicate that, ideally, the thrust-line of an arch should be identical with the locus of the centroids of its cross-section, so that the bending moment everywhere is zero. Where dead load is dominant the free bending moment would, therefore, determine the shape of the arch, and the necessary abutment condition is that horizontal deflexion is prevented.

In his paper on the elastic arch (1898) Young applies Castigliano's least work principle (with acknowledgement) but he seems unaware of the contributions of Winkler and Fränkel. However, the application of the energy method on the continent of Europe was commonplace by this time.

Stiffened suspension bridges

Fränkel's energy analysis of the elastic behaviour of suspension bridges, which seemed to escape the attention of Levy (1886) (who adopted an approach from first principles), is preceded by a diagram (Fig. 54; being

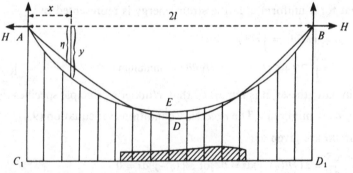

Fig. 54. From Fränkel (1882).

Fig. 5 of Fränkel's article). In the diagram H is the horizontal component of tension in the elastic cable, caused by live loading applied to the deck, and indicated by a shaded area (dead load is assumed to cause no bending of the deck because the suspension bridge concept is that the self-weight (dead load) is transferred entirely to the cable system). By ingenious analogy with an arch, the difference between the actual shape of the cable system AEB and the curve ADB is assumed to be proportional to the bending moment in the deck girders caused by the live load. At a distance x from the origin A, the ordinate of the former is y and that of the latter is η. The tangent of the curve AEB subtends angle τ to the horizontal axis at any point along the curve, distant s from the origin A. The elastic energy of the cable, due to the live load, is written:

$$\int_0^{2l} \frac{H \sec \tau}{2} \frac{H \sec \tau}{EF} \, ds = \int_0^{2l} \frac{(H \sec \tau)^2}{2EF} \, ds \qquad [(9.7)]$$

where $2l$ is the span; E the modulus of elasticity; and F the cross-sectional area of the cables or chains; while the strain energy of the deck structure (flexural rigidity $E_1 J_1$), due to the live load, is:

$$\frac{1}{2} \int_0^{2l} \frac{M^2}{E_1 J_1} \, dx \qquad [(9.8)]$$

Thus the total strain energy, due to the live load, is:

$$A = \frac{1}{2} \int_0^{2l} \frac{(H \sec \tau)^2}{EF} \, ds + \frac{1}{2} \int_0^{2l} \frac{M^2 \, dx}{E_1 J_1} \qquad [(9.9)]$$

Then, by means of the arch analogy, Fränkel says that by definition $M = H(\eta - y)$ and writes as his equation (18):

$$A = \frac{1}{2} \int_0^{2l} \frac{(H \sec \tau)^2}{EF} \, ds + \frac{1}{2} \int_0^{2l} \frac{H^2(\eta - y)^2 \, dx}{E_1 J_1} \qquad [(9.10)]$$

and if

$$\eta = B + Cx + \frac{f(x)}{H} \qquad [(9.11)]$$

then

$$(\eta - y)H = BH + CHx + f(x) - Hy \qquad [(9.12)]$$

By the condition that $\eta = 0$ at $x = 0$ and at $x = 2l$:

$$0 = B + \frac{f(0)}{H}$$

and

$$\qquad [(9.13)]$$

$$0 = B + Cl + \frac{f(2l)}{H}$$

therefore

$$B = -\frac{f(0)}{H}$$

and

$$\qquad [(9.14)]$$

$$C = \frac{f(0)}{Hl} - \frac{f(2l)}{Hl}$$

Here it should be noted that Fränkel made an error in substituting for $x = 2l$ in equation (9.11), for it seems as though the second of equations (9.13) should, in fact, read:

$$0 = B + 2Cl + \frac{f(2l)}{H}$$

Fränkel substituted for $H(\eta-y)$ in equation (9.10) using the expression for η of equation (9.11) including the values of B and C as in equations (9.14) to obtain:

$$2A = \int_0^{2l} \frac{(H\sec\tau)^2}{EF}\,ds + \int_0^{2l}\left\{-f(0)+\frac{f(0)}{l}x-\frac{f(2l)}{l}x+f(x)-Hy\right\}^2\frac{dx}{E_1J_1}$$

[(9.15)]

and then by his least work principle:

$$\frac{dA}{dH} = 0 = \int_0^{2l}\frac{H\sec^2\tau}{EF}\,ds - \int_0^{2l}\left\{-f(0)+\frac{f(0)}{l}x-\frac{f(2l)}{l}x+f(x)-Hy\right\}\frac{y\,dx}{E_1J_1}$$

[(9.16)]

so that:

$$\int_0^{2l}\frac{H\sec^2\tau}{EF}\,ds = \int_0^{2l}H(\eta-y)y\frac{dx}{E_1J_1}$$

[(9.17)]

But the bending moment of the deck structure as a simply supported beam is $M' = H\eta$ due to the live loading, and Fränkel writes (his equation (19)):

$$\int_0^{2l}\frac{H\sec^2\tau}{EF}\,ds = \int_0^{2l}\frac{(M'-Hy)y\,dx}{E_1J_1}$$

[(9.18)]

Next he specifies y in the parabolic form:

$$y = \frac{hx(2l-x)}{l^2}$$

[(9.19)]

where h is the dip at the centre of the cable. Also, he expresses the cross-sectional area F of the cable at any point as $F = F_0(ds/dx)$ and specifies E_1J_1 as constant, hence:

$$\int_0^{2l}\frac{H\sec^2\tau}{EF}\,ds = \frac{H}{EF_0}\int_0^{2l}\left\{1+\left(\frac{dy}{dx}\right)^2\right\}dx = \frac{1}{E_1J_1}\int_0^{2l}(M'-Hy)dx$$ [(9.20)]

or

$$\frac{H2l}{EF_0}\left\{1+\frac{4}{3}\frac{h^2}{l^2}+\frac{h^2}{4l^2}\right\} = \frac{1}{E_1J_1}\int_0^{2l}(M'-Hy)dx$$

[(9.21)]

If, for example, the live load is a single concentrated load P, distant a from C (Fig. 54), then

when $x < a$:

$$M' = \frac{P(l-a)}{2l}x$$

and when $x > a$:

$$M' = \frac{P(l-a)}{2l}-P(x-a)$$

[(9.22)]

so that substitution in equation (9.21) gives:

$$H = \frac{5}{64} P \frac{a(8l^3 - 4la^2 + a^3)}{l^3 h \left\{ 1 + \frac{15}{8} \frac{E_1 J_1}{EF_0 h^2} \left(1 + \frac{4h^2}{3 l^2} \right) \right\}} \qquad [(9.23)]$$

(Fränkel refers here to 'Müller-Breslau, Theorie der Durch einen Balken versteiften Kette auf S. 61 des Jahrg. 1881 dieser Zeitschrift').

Having found H, the bending moment in the deck girders due to live load, p may be found because $M = H(\eta - y)$. Comparison of equation (9.23) with the results of Levy's later investigation (Chapter 3), using first principles, is readily accomplished if the cables are inextensible. Then $EF_0 = \infty$ and H and M are independent of $E_1 J_1$ and in agreement with Levy's theory.

Fränkel's suspension bridge analysis by the elastic theory, using the so-called least work principle, avoids much tedious (though instructive) detail, as study of the investigations of both Levy and Du Bois indicates, in specifying the necessary condition of compatibility of deflexion of the cables and deck. It is especially interesting that he seemed to be so concerned with demonstrating the versatility of a novel (general) principle that he overlooked the possibility of priority in respect of an elastic theory of suspension bridge behaviour.

In passing, it is worthwhile to recall the contents of the note, appended to Cotterill's article (1865*b*), on the equilibrium of arched ribs by his energy method (Chapter 8). He says in that note: 'The general problem of the stiffened suspension bridge is a particular case of this more general problem. The elasticity of the chains can be taken into account by estimating the work done in them in terms of H_0'.' (The horizontal component of tension in the chains, which Cotterill required to be a minimum is H_0'.) In fact, Cotterill chose the suspension bridge to illustrate the application of his principle in his earlier article (1865*a*). Having neglected the elasticities of chains, bars and piers of the complete structure, he arrived at results which, he says, 'agree exactly with those obtained by a writer in *Civil Engineer and Architect's Journal* for 1860, and differ slightly from those given by Professor Rankine in his work on applied mechanics'. (The publication to which he refers is considered in Chapter 3). Cotterill, moreover, mentions in general terms the possibility of taking account of the elasticities of suspension chains *and* bars in the analysis.

Müller-Breslau is notable among those who followed Winkler and Fränkel in exploitation of the least work principle, with particular acknowledgement to Castigliano and application to arches and suspension bridges.

Unfortunately, the later, more precise, theory of suspension bridge behaviour, namely the deflexion theory which was due initially to Melan (1888), seemed to defy a convenient approach by energy. It is necessary only to compare the effort expended by Levy with that involved in the derivation of Fränkel's energy approach to the elastic theory, to appreciate the expectations generated by the energy device. It is, however, worth

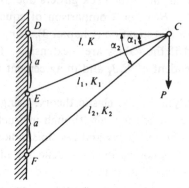

Fig. 55. From Fränkel (1882).

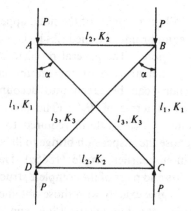

Fig. 56. From Fränkel (1882).

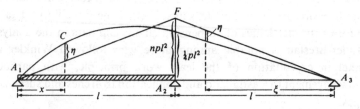

Fig. 57. From Fränkel (1882).

emphasising that the elastic theory remains relevant to suspension bridges if deck girder stiffness is great.

Fränkel finally demonstrates the use of the energy method for bar frameworks and continuous beams, with reference to the examples shown in Figs. 55, 56 and 57, where K is bar force and η (Fig. 57) is the resultant bending moment. In the analysis η is expressed as a function of the reaction of the intermediate support, npl, where n is a fraction of the distributed load pl.

10

Aspects of the further development of theory of structures

After the theory of statically-indeterminate frameworks was established in Europe *c.* 1875, theory of structures advanced rapidly, especially by virtue of the property of a linear relationship between 'cause' and 'effect', which characterised engineering structures and which afforded the principle of superposition (Chapter 3) and the reciprocal theorem (Chapter 5). Dominant among the contributors to these advances were Mohr and Müller-Breslau in Germany, and much of this chapter is concerned with them and their work. Also, the Italian railway engineer, Crotti, deserves special mention for his unique contribution to the development of a general theory of elastic structures.

Mohr and Müller-Breslau

Mohr, born in 1835, was some sixteen years older than Müller-Breslau but nevertheless there seems to have been antagonism and rivalry between them, to judge, especially, from published comment by the latter, which is illuminating in various respects and is therefore included in this chapter. Müller-Breslau died in 1925, only seven years after Mohr. His later work was arranged for publication by his son who was a professor at the Breslau Polytechnikum. The final edition of part of his monumental work *Graphische Statik der Baukonstruktionen* anticipated important developments in theory of structures in the present century.

At the age of thirty-three, Mohr became professor of engineering mechanics at Stuttgart, having spent the early years of his career in railway construction in common with many distinguished civil engineers of the nineteenth century. Then in 1873 he became professor of structural engineering at the Dresden Polytechnikum where he remained until 1900. He had the reputation of being an excellent teacher and is still widely known for his elegant graphical methods, especially the circle diagrams for complex stress and strain. His equally elegant analytical approach to the analysis of statically-indeterminate frameworks seems to have been first published in English by Swain (1883) who had made known Mohr's graphical construction for earth pressures (1882). (Indeed, Swain contributed greatly to the introduction of important German works to the English speaking world at this time, including those of Winkler and the analysis of two-pin framed arches by Schultze in 1865 and Fränkel in 1875.)

Müller-Breslau was docent and then professor at the Hannover Polytechnikum for five years before he succeeded to the professorial chair – of bridge engineering – at Berlin in 1888 on the death of Winkler. While at Hannover he published his celebrated book *Die neueren Methoden* (1886b). He was probably unique among the giants of nineteenth-century theory of structures, in making such important contributions to the design and testing of aircraft and rigid airship structures that he received the Iron Cross during the First World War.

The ill-feeling between Mohr and Müller-Breslau seems to have arisen from their respective attitudes to the work, in theory of structures, of Maxwell and Castigliano. Müller-Breslau, the younger man, seemed to be thoroughly up to date with regard to his appreciation of contemporary research in Europe. He readily (perhaps too readily for Mohr) acknowledged Maxwell's priority with regard to the principle involved in the particular approach to the analysis of statically-indeterminate frameworks, which Mohr discovered in 1874. But there seem good grounds for doubting the assertion by Niles (1950) that Müller-Breslau regarded Mohr as a 'mere copyist' since he seemed to give generous acknowledgement to Mohr whenever appropriate.

Mohr, on the other hand, seemed to care little for the work of his contemporaries. His first reference to the work of Maxwell and Castigliano seems to have been in an article published in 1885 (Chapter 5). In a review of then current technical literature in his book of collected topics in technical mechanics (1906) he mentions the article of 1885, claiming that it gives a summary of previous articles about the advantages of the principle of virtual work, including Clapeyron's derivation of his theorem, together with discussion of the alternative methods of Maxwell and Castigliano. He emphasises that his own approach, using virtual work, affords the solutions to all problems in theory of structures! It is, moreover, significant that in the book (1906) he credits Menabrea as the author of the principle of least work, and Castigliano merely with the theorem that the partial derivative of the strain energy of a linearly elastic structure, with respect to an external force, gives the deflexion in the line of action of that force.

Müller-Breslau provided an interesting and revealing commentary on Mohr's attitude in a footnote of *Die neueren Methoden* (1886*b*, p. 188), where he says:

We cannot avoid mentioning at this point the derogatory verdict of Professor Mohr in his article (1885) concerning Castigliano's principles and which is based on the assertion that there is an implicit error in those principles (although they will provide correct results) in that the statically-indeterminate values X are treated as independent variables of the work A so that it follows ultimately that the principle $\dfrac{dA}{dP} = \delta$ does not produce a profitable result for a statically-indeterminate framework, even though it is of itself a correct statement. Mohr overlooks the fact that if A is taken for the whole framework and X a function of the loading, it is only necessary to consider the conditions that X must fulfil, in performing the differentiation, to see at once that the value of δ is independent of the differential quotients of the quantities X which depend on P. It also seems to have escaped Mr Mohr's attention that the principle in question holds true for any individual part of the framework, such as, for example, the statically-determinate 'main network' on which the forces in redundants act as external loads... In this way $\dfrac{dA}{dP}$ can be used to obtain strain compatibility equations of elasticity and, subsequently, the deflexions.

Müller-Breslau's own attitude is revealed in his historical survey of principles relating to the analysis of statically-indeterminate frameworks, near the end of his book of 1886. He begins by remarking that the first use of energy devices for elastic structures is to be found in the work of Clapeyron who used the general condition for equilibrium based on virtual work 'as developed by Navier'. Also, he noted that Clapeyron used elastic displacements as virtual displacements to derive what (he said) Lamé described as 'Clapeyron's Law'. Then, credit is given to Maxwell for his method of analysing statically-indeterminate frameworks (but Navier's method is not mentioned). Müller-Breslau continues by saying that the first comprehensive account of the theory of statically-indeterminate frameworks was based on the principle of virtual work and was given by Mohr (1874*a*, 1874*b*) and included derivation of 'Maxwell's theorem' and influence lines for deflexions of joints of structures. He said that Mohr also obtained the elastic line of a beam and the deflexion pattern of a framework by means of the link (funicular) polygon.

Müller-Breslau thought that mechanics of materials and, in particular, the theory of statically-indeterminate structures was advanced extensively by the 'brilliant work of the Italian engineer Castigliano who tragically died so young'. (Indeed, in his *Graphische Statik* (1892, vol. 2), he states his belief that Castigliano's contribution is greater than that of Maxwell, which, he said, was limited to consideration of a plane framework.) He quotes especially the principle of least work, saying that the principle 'had been stated by Menabrea in an earlier work (1858) and independently by Fränkel (1882)'. In addition he wrote:

It is also noteworthy that Daniel Bernoulli also formulated a principle of least energy of deformation of straight bars and this was mentioned in correspondence with Euler. Euler made use of this in his famous works *Methodus inveniendi curvas maximi minimive proprietate gaudentes* and *De curvis elasticis*, in which he starts the investigation of the elastic line of a uniform straight bar: 'ut inter omnes curvas ejusdem longitudinis, quae non solum per puncta *A* et *B* transeant, sed etiam in his punctis a rectis positione datis tangantur, definiatur ea in qua sit valor hujus expressionis $\int \frac{ds}{R^2}$ minimus'. Here d*s* is an element of the arc and *R* the radius of curvature. Since $\frac{1}{R} = \frac{M}{EJ}$, it may be concluded, *EJ* being constant, that $\int \frac{M^2}{EJ} ds$ is a minimum. Bernoulli called the integral: $\int \frac{ds}{R^2}$ the '*vis potentialis*'.

Mohr's method for deflexion of frameworks

Mohr's dexterity and insight into the theory of beams in bending led him to an elegant device for calculating the deflexions of the joints of a pin-jointed bar framework (bridge truss) which he included in the final section of his article 'Beitrag zur Theorie des Fachwerks' (1875). He showed that the vertical deflexion of the joints of a chord (boom) of such a truss, due to the change in length of any bar of the chord, may be represented by the bending moment diagram for a simply supported beam, due to a concentrated load. The details of the method are shown in Fig.

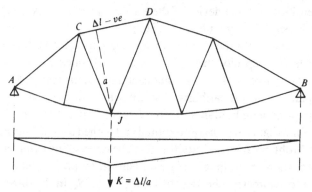

Fig. 58. After Mohr (1875).

58 (Fig. 30 of Mohr's article 1875) with reference to the framed girder shown. In order to determine the deflexion due to bar CD which extends by amount Δl, due to the loading applied to the girder, and regarding all other bars as inextensible (or incompressible), it is necessary to consider the effect of the rotation about J of the two sections of the girder adjacent to CD through the small angle $\Delta l/a$ (if a is the distance from J to CD as shown). The consequent vertical deflexions of points along the chords of the girder may be shown by simple geometry to be given by the values of the bending moment, in a simply supported beam of the same length as the girder, due to a load numerically equal to $\Delta l/a$, at a point corresponding to J. Mohr also showed that the effect of the small change in length of any diagonal bar on the deflexions of the girder, may be described in a similar manner. Thus, by sketching such a diagram for each bar and having constructed the influence line for the change in length of each bar, vertical deflexions of the girder, due to any specified loading, may be determined by a simple numerical process. (Winkler showed (1881b) that the deflexions of the joints of one chord of a girder may be determined simply by considering the behaviour of the members of that chord.)

Mohr's elastic centre of frameworks

A particularly elegant contribution to the analysis of statically-indeterminate structures was made by Mohr in 1881. He considered the analysis of framed arches with three degrees of statical indeterminacy, as for an encastré arched rib, and showed how the three simultaneous equations of compatibility of deformation, relating the statically-indeterminate forces, can be transformed to three independent equations, each containing only one quantity relating to those unknowns. Indeed, it was essentially the transformation to the relevant normal coordinates and it may be identified with the device of the elastic centre which was defined originally, it is believed, by Culmann (1866).

Thus, in his article 'On the theory of framed arches' (1881), Mohr sets out to show how to specify the redundants of a framed arch which is symmetrical about a vertical axis, so that each redundant features in only one of the equations of compatibility of strain, which are necessary for the purpose of finding the redundants. With reference to the structure shown in Fig. 59 (Mohr's Fig. 1), which is pin-jointed and pinned to rigid abutments at A, B, C, D, Mohr considers the abutments to be replaced by a framework whose bars are rigid or incapable of strain (shown by broken lines) and, having regard to symmetry, chooses the forces S_1, S_2, S_3, in the three bars shown, as the three redundants of the total ring framework. He specifies

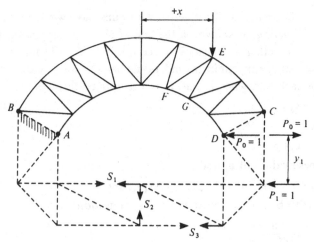

Fig. 59. From Mohr (1881).

that the distance of the lines of action of S_1 and S_3 from the line *AD* are y_1 and y_3 respectively, and are positive or negative depending upon whether they are below or above *AD*. With reference to his earlier articles (1874a, 1874b) he gives the strain compatibility equations in terms of the unknown redundants as:

(1)
$$0 = \Sigma u_1 Gr + \Sigma u_1 l\delta t + S_1 \Sigma u_1{}^2 r + S_2 \Sigma u_1 u_2 r + S_3 \Sigma u_1 u_3 r$$
$$0 = \Sigma u_2 Gr + \Sigma u_2 l\delta t + S_1 \Sigma u_1 u_2 r + S_2 \Sigma u_2{}^2 r + S_3 \Sigma u_2 u_3 r$$
$$0 = \Sigma u_3 Gr + \Sigma u_3 l\delta t + S_1 \Sigma u_1 u_3 r + S_2 \Sigma u_2 u_3 r + S_3 \Sigma u_3{}^2 r$$

[(10.1)]

where the summations include the bars of the actual framework only, since the temperature effects and elasticities, $r = l/EF$, of those of the imaginary foundation framework are zero; *l* being the length; *EF* the product of modulus of elasticity and cross-sectional area; and δt is any temperature variation. The quantities u_1, u_2, u_3 are the forces in the bars of the actual framework due to $S_1 = 1$, $S_2 = 1$, $S_3 = 1$ respectively; and *G* is the force in a bar due to any specified external loading.

From considerations of symmetry of the framework considered, Mohr notes that $\Sigma u_1 u_2 r$ and $\Sigma u_2 u_3 r$ are zero: however, $\Sigma u_1 u_3 r$ is not zero but by judicious choice of y_1 and y_3 (the 'levels' of S_1 and S_3 respectively) it can become zero. It is the method of making that judicious choice which completes Mohr's task because then equations (10.1) are reduced to three independent equations in the unknowns S_1, S_2, S_3. Mohr proceeded by specifying u_0 and u as the typical tractions which appear in the framework when the redundants are removed and it is loaded by $P_0 = 1$ through *D*, to the left (as shown in Fig. 59), together with a unit clockwise couple. Also, a force $P_1 = 1$ acting, as shown, in the line of S_1 can be expressed as the sum of $P_0 = 1$, together with a clockwise couple $y_1 P_0$ ($P_0 = 1$) due to which, bar forces u_0 and $y_1 u$ appear in any individual bar of the actual framework.

Thus, for each bar, when a load $P_0 = 1$ is applied through D:

(2) $u_1 = u_0 + y_1 u$ [(10.2)]

and similarly:

(3) $u_3 = u_0 + y_3 u$ [(10.3)]

Now for these values of u_1 and u_3:

$$\Sigma u_1 u_3 r = \Sigma u_0{}^2 r + (y_1 + y_3) \Sigma u u_0 r + y_1 y_3 \Sigma u^2 r$$ [(10.4)]

and if this is to be zero, y_1 must satisfy the condition:

(4) $y_1 = -\dfrac{(\Sigma u_0{}^2 r + y_3 \Sigma u u_0 r)}{(\Sigma u u_0 r + y_3 \Sigma u^2 r)}$ [(10.5)]

Among the infinite number of pairs of values of y_1 and y_3 which satisfy this condition, Mohr asserts that those in which either y_1 or y_3 is very large have some advantage. If y_3 is very large in relation to all other quantities, then by equation (10.5):

(5) $y_1 = -\dfrac{\Sigma u_0 u r}{\Sigma u^2 r}$ [(10.6)]

and

(6) $u_3 = y_3 u$ [(10.7)]

Finally, then, subject to this condition and having regard to the zeros observed by symmetry, equations (10.1) become:

(7)
$$S_1 = -\frac{(\Sigma u_1 Gr + \Sigma u_1 l \delta t)}{\Sigma u_1{}^2 r}$$

$$S_2 = -\frac{\Sigma u_2 Gr + \Sigma u_2 l \delta t}{\Sigma u_2{}^2 r}$$ [(10.8)]

$$y_3 S_3 = M = -\frac{\Sigma u Gr + \Sigma u l \delta t}{\Sigma u^2 r}$$

The product $y_3 S_3 = M$ is the moment of a small force S_3 at a large distance from the line AD or, quite simply, the couple due to S_3. Thus, the couple acting at the right-hand side of the framed arch in the absence of S_1 and S_2 is clockwise or anti-clockwise, depending on whether M is positive or negative.

By the highly ingenious method of reasoning described, Mohr had, in effect, located the elastic centre of a symmetrical framed arch. Thus (as shown in Fig. 60) forces S_1 and S_2, acting through a point on the vertical

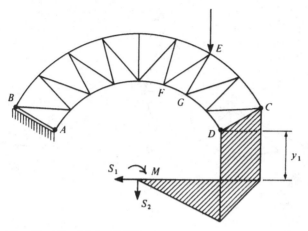

Fig. 60. After Mohr (1881).

axis of symmetry at a distance y_1 below the level of D, together with a couple M, are equivalent to the three forces at the abutment CD, which provide complete restraint there. In fact, it is as though they acted on the structure through the medium of a rigid arm, as shown, which is simply another way of regarding the foundation structure of rigid bars, which Mohr introduced to deduce his method. He showed that y_1 is likely to be negative, so that the elastic centre is above, rather than below, D. Indeed, it is because of the location of rigid foundation structure, shown by him, that the article (at first sight) is unfamiliar, though his is the obvious choice without prior knowledge.

Mohr proceeded to describe an elegant device for the expeditious determination of the effect of external loads, which is compatible with the concepts he introduced for his earlier derivation. Such devices for facilitating computation are typical of the development of structural analysis in Germany, especially by Mohr, Müller-Breslau and Winkler; they are the result of profound appreciation of principles and close attention to detail.

Müller-Breslau's methods

Müller-Breslau also gave a lot of attention to simplifying the simultaneous equations of statically-indeterminate structures, by choosing the forces representing the redundants in such a way that, at best, each of those forces appeared in one equation only. Like Mohr, he seems to have concentrated at first on framed arches for this purpose, as is evident in his comprehensive article (1884c), 'Vereinfachung der Theorie der

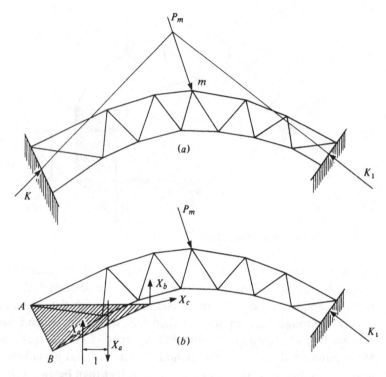

Fig. 61. From Müller-Breslau (1892).

Statisch unbestimmten Bogenträger'. He is, indeed, sometimes credited with priority in the solution of the problem (for example by Pippard & Baker (1957)). It is interesting to contrast his approach with that of Mohr, which comparison of Fig. 61 (from Müller-Breslau's *Graphische Statik*, 1892, vol. 2) with Fig. 59 indicates. It was, moreover, not long before Müller-Breslau simplified conceptual thought in the analysis of statically-indeterminate structures by introducing the principle of influence or flexibility coefficients (probably first used by the mathematician Clebsch, as noted in Chapter 5, but apparently unknown to engineers). Thus, coefficients of the forces in redundants, such as $\Sigma\, u_i u_j r$ in Mohr's equations of compatibility of strain, become, in Müller-Breslau's notation, simply a_{ij} or δ_{ij}. Moreover, the method used for their determination then ceases to overshadow the analytical process.

In his earlier book (1886b), Müller-Breslau expresses the equations of compatibility of strain in terms of forces in redundants Z and deformations c caused by external loading, as:

$$a_{11}Z_1 + a_{12}Z_2 + a_{13}Z_3 + \ldots + a_{1i}Z_i = c_1$$
$$a_{21}Z_1 + a_{22}Z_2 + a_{23}Z_3 + \ldots + a_{2i}Z_i = c_2$$
$$\vdots$$
$$a_{i1}Z_1 + a_{i2}Z_2 + a_{i3}Z_3 + \ldots + a_{ii}Z_i = c_i$$

[(10.9)]

while in his later book (1892) he uses the notation:

$$\delta_{11}X_1 + \delta_{12}X_2 + \delta_{13}X_3 + \ldots + \delta_{1i}X_i + \ldots + \delta_{1n}X_n = Z_1 \qquad [(10.10)]$$

Thus, not only is δ used instead of a, but Z replaces c to denote deformations caused by loads (and temperature effects) and instead of i redundants with forces Z, there are n redundants with forces X. Moreover, he specifically defines Z_i as $\Sigma\, P_m \delta_{mi}$, using the symbol P to denote external load, and introduces a term for temperature strain. Also, he defines the $n \times n$ matrix of the flexibility coefficients and observes that $\delta_{ij} = \delta_{ji}$ (in accordance with the reciprocal theorem) so that the matrix is symmetrical about the leading diagonal (*Hauptdiagonale*).

Müller-Breslau gives the inversion of the equations (10.9) of which the ith is:

$$X_i = \beta_{i1}Z_1 + \beta_{i2}Z_2 + \beta_{i3}Z_3 + \ldots + \beta_{ii}Z_i + \ldots + \beta_{in}Z_n \qquad [(10.11)]$$

in the revised notation, and notes that again, the coefficients β (stiffness coefficients) of the deformations Z are embodied in a symmetric square matrix.

Although Müller-Breslau generalised framework analysis by taking the fullest advantage of the properties of linearity and enabled the subject to be placed above controversy over the relative merits of virtual work, Clapeyron's theorem or Castigliano's theorems for detailed computations, it is apparent that the rigidly-jointed portal framework was not then a matter of primary concern. Thus, he apparently did not pursue the implications of equations (10.11), no doubt because, for lattice frameworks, they appeared of academic interest, being concerned with displacements in the lines of redundants rather than of joints. The fact that his early (1886*b*) accounts of the analysis of portal frameworks are confined to simple forms and his use of Castigliano's least work method (and later the Maxwell–Mohr method) tend to confirm this impression. Like his contemporaries he used only the approach whereby the forces in redundants are the unknowns. The alternative was to await the attention of Ostenfeld early in the present century.

Müller-Breslau investigated the conditions necessary to render coefficients, such as δ_{ij} of the chosen basic (statically-determinate) system, zero for a variety of structures in addition to framed arches. For example Fig. 62 shows a structure for which he chose (1892) two components of reaction

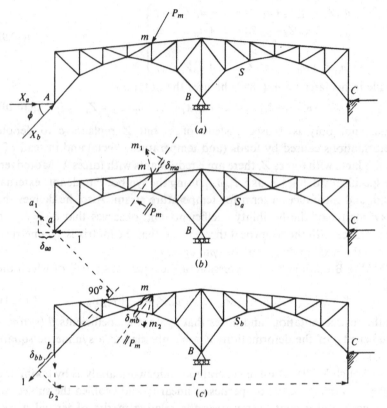

Fig. 62. From Müller-Breslau (1892).

at A as redundants and, denoting them by X_a and X_b, proceeded to find their lines of action so that:

$$X_a = \frac{Z_a}{\delta_{aa}} \quad \text{and} \quad X_b = \frac{Z_b}{\delta_{bb}}$$ [(10.12)]

Reduction method for deflexions

Curiously, a very elegant device for finding the deflexion of a particular joint of a complicated bar structure, which Müller-Breslau gave in his *Die neueren Methoden* (1886b), seemed to be unknown in Britain until comparatively recently. It is illustrated in Fig. 13 of that book (shown here in Fig. 63). The problem is to find the deflexion at m, the mid-point of the lower chord of the continuous framed girder $C_0 C_1 C_2 C_3$ shown, due to any specified live loading. It is necessary, first of all, to calculate the forces in the bars of the complete structure, due to the specified loading (by the

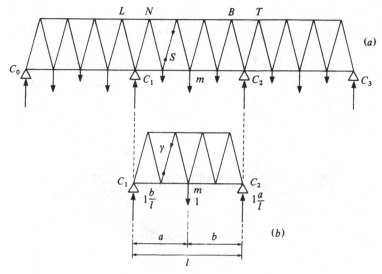

Fig. 63. From Müller-Breslau (1886b).

Maxwell–Mohr method which is appropriate for such structures and depends upon equations relating the forces in the chosen redundants and the loading, based upon compatibility of elastic displacements) and then to find the corresponding changes in length of all of the bars which are relevant. Thus, it is the span $C_1 C_2$ alone which is relevant, as shown in Fig. 63(b), with unit force applied at m and bar forces S', because:

$$1\delta_m = \sum_{C_1}^{C_2} S' \Delta_s \qquad\qquad [(10.13)]$$

That is, the actual small displacements δ_m of m and Δ_s of a bar, found by the analysis of the complete structure for the specified loading, are a possible compatible set with regard to any system of forces in equilibrium in the structure $C_1 C_2$ by itself, in particular that caused by unit load at m, whereby δ_m is determined by equation (10.13). It is likely that this device (*Reducktionsätz*) was well known to Mohr as a protagonist of the principle of virtual work in theory of structures.

Müller-Breslau's principle

The contribution to theory of structures for which Müller-Breslau's name is perhaps best known nowadays, is concerned with influence lines for statically-indeterminate structures, a topic which clearly attracted a great deal of his attention. In 1883 he published an article on influence lines for continuous structures with three supports, which was repeated in

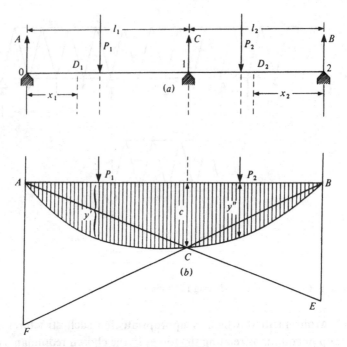

Fig. 64. From Müller-Breslau (1884*b*).

substance in 1884 (also in *Die neueren Methoden*, 1886*b*, his Fig. 97). The important feature is apparent from the outset. Thus, his Figs. 1 and 3 of the second article (1884) are shown in Fig. 64(*a*) and (*b*) respectively; the former shows a beam 012 on three simple supports at 0, 1 and 2, with live loads P_1 and P_2. In Fig. 64(*b*) the curve ACB, with ordinates y' in the line of action of P_1, and y'' in the line of action of P_2, represents the effect of introducing a small arbitrary displacement c in the line of action of the reaction C of the intermediate support to those loads. On the basis of the reciprocal theorem, Müller-Breslau shows that the reaction C is given by:

$$C = \frac{1}{c}(P_1 y' + P_2 y'') \qquad\qquad [(10.14)]$$

and that the curve ACB represents the influence line for C. Thus the influence line for the force in any element of a structure is represented by the deflexion pattern produced by the introduction of a small deflexion in the line of action of the force in the chosen element. This is usually called 'Müller-Breslau's principle'. It is clearly related to Mohr's theorem concerning influence lines for deflexion of linearly elastic structures (Chapter 5) but it is more general in that it is also relevant to statically-determinate structures. (Other structures considered by Müller-Breslau in his article include bar frameworks on three simple supports.)

Müller-Breslau's contribution to the analysis of frameworks with rigid joints, was concerned especially with secondary stresses in bridge trusses (1886*a*), as noted in Chapter 11, though he was among the first (1886) to analyse rigidly-jointed, portal frameworks.

Fränkel and Winkler

Fränkel and Winkler also made profound contributions to the further development of structural analysis in the nineteenth century. Fränkel, born in 1811, was the senior of the distinguished German school which established leadership in theory of structures during the century, while Winkler was Mohr's contemporary; they were both born in 1835, some sixteen years earlier than Müller-Breslau. Both Fränkel and Winkler were railway engineers at first; the former became docent and professor at the newly founded Polytechnikum of Dresden (where he was subsequently joined by Mohr), the latter became professor of bridge and railway engineering at Prague, then Vienna and finally Berlin, where he died in 1877 at the age of fifty-three and was succeeded by Müller-Breslau.

Their close collaboration is revealed in the comprehensive treatment of influence lines by Fränkel (1876*b*), following Winkler's discovery of the concept (1868) almost simultaneously with Mohr (Chapter 5), and by Fränkel referring the manuscript of his celebrated paper (1882) on a proposed energy principle, to Winkler for the latter's approval. (Fränkel's action resulted in Winkler's advice, as noted in Chapter 8, that he had been anticipated in principle by Castigliano several years earlier.) The outstanding merit of that paper is the originality and variety of the examples of application of the principle, including a novel approach to the elastic theory of suspension bridges (Chapter 9). Fränkel is also widely believed to have originated the reduction method with virtual work for calculating deflexions of structures in the manner described by Müller-Breslau (1886*b*).

Both contributed extensively to application of the new analytical devices to bridge design and both contributed to the analysis of secondary stresses in bridge trusses – Fränkel, by devising an extensometer (1881) and *in situ* measurements on railway bridges; and Winkler by an analytical method (Chapter 11). But Winkler is probably best known nowadays, for his theory of bending of thick curved elements, including crane hooks (1858), theory of beams on elastic foundations and continuous beams.

Crotti

An important later contribution is due to Crotti, one aspect of which is considered in Chapter 8. In the introductory chapter of his celebrated monograph (1888), he reviews the development of theory of elasticity and mentions especially, Galileo, Hooke, Mariotte, Coulomb and Navier. Thus, he credits Coulomb with having correctly determined the position of the neutral axis of an elastic beam, and remarks on the error, in that respect, in Navier's course of lectures of 1819, which was corrected in his lithographed notes of 1824. But Crotti subscribes to the belief that modern theory of elasticity is due to Navier. Having turned his attention at some length to Green and his theorem and its application to energy devices, he pays tribute to Saint-Venant and then to Clapeyron for his theorem of the three moments for continuous beams. He accords warm acclaim to Lamé's *Leçons* (1852) and notes an indirect similarity of approach to that of Green (1839) in some important respects. The 'brilliant' contribution, due to the genius of Helmoltz (1847), is noted with regard to general application of the doctrine of energy and the revelation of the principle of duality 'which always exists between displacement and force' (Chapter 7).

Crotti's approach (1888) to the theory of elastic structures is abstract in nature and he does not concern himself with application to other than trivial problems, to the detriment of the appeal of his work to engineers. His derivations are based on energy functions, including elastic energy (work) which he denotes by the symbol L. He apparently neither mentions nor uses the principle of virtual work, however. Denoting force by F and displacement in its line of action by f, he shows that for any two forces F_r and F_s acting upon an elastic solid or structure, causing displacements f_r and f_s respectively:

$$\frac{\mathrm{d}L}{\mathrm{d}f_r} = F_r; \quad \frac{\mathrm{d}L}{\mathrm{d}f_s} = F_s \qquad\qquad [(10.15)]$$

by partial differentiation. By partially differentiating the first with respect to f_s, and the second with respect to f_r, he concludes:

$$\frac{\mathrm{d}F_r}{\mathrm{d}f_s} = \frac{\mathrm{d}F_s}{\mathrm{d}f_r} \qquad\qquad [(10.16)]$$

which is tantamount to a statement of the reciprocal theorem. Indeed, Crotti later uses the designation a_{sr} for $\mathrm{d}F_r/\mathrm{d}f_s$, and a_{rs} for $\mathrm{d}F_s/\mathrm{d}f_r$ and he expresses forces and loads in terms of linearly elastic displacements as:

$$
\left.
\begin{aligned}
F_1 &= a_{11} f_1 + a_{12} f_2 + \ldots + a_{1n} f_n \\
F_2 &= a_{21} f_1 + a_{22} f_2 + \ldots + a_{2n} f_n \\
&\ \ \vdots \\
F_n &= a_{n1} f_1 + a_{n2} f_2 + \ldots + a_{nn} f_n
\end{aligned}
\right\}
\qquad [(10.17)]
$$

Then he expresses the displacements in terms of the forces as:

$$
\begin{aligned}
f_1 &= A_{11} F_1 + A_{12} F_2 + \ldots \\
f_2 &= A_{21} F_1 + A_{22} F_2 + \ldots
\end{aligned}
\qquad [(10.18)]
$$

where $A_{rs} = A_{sr}$ (that is, the flexibility coefficients (A_{ij}), as well as the stiffness coefficients (a_{ij}), possess the property of reciprocity).

Crotti also showed that for a quantity $\delta\lambda = \Sigma f \delta F : d\lambda/dF_r = f_r : d\lambda/dF_s = f_s : \ldots$, and thence that $df_r/dF_s = dF_s/dF_r$.

Explicit recognition and demonstration of the principle of duality, with regard to the forces and displacements of elastic systems, is a unique feature of Crotti's work, in comparison with that of his contemporaries.

Weyrauch

Finally, the contributions of Weyrauch (1845–1917), professor at the Polytechnikum of Stuttgart (after his experience as a railway engineer) are significant. He wrote several important and comprehensive books on theory· of elasticity (1884, 1885); bridge structures, including continuous beams (1873, 1888); and elastic arch structures (1896). He is, moreover, widely credited for his early application of influence lines to bridge design (including the terminology). At the end of the century he wrote an article (published 1901) on the history of bridge building, in which he calculates the limiting span of a steel suspension bridge as 3730 m, assuming a dip of one-third of the span and a safety factor of three.

Two of his books (1887, 1888) are of unique interest, being devoted to the analysis and design of continuous structures which retain statical determinacy, including a very detailed account of methods of analysis. He seems intent on demonstrating that a judicious approach to the economic design of bridge structures renders statical indeterminacy unnecessary, and that in circumstances where it is allowed to occur (such as the so-called 'double Warren girder'), use of the device of superposing two symmetrical statically-determinate systems suffices for design purposes. Indeed, he developed statical determinacy to a remarkable extent (see also Chapter 4).

Weyrauch was especially interested in the doctrine of energy, including energy derivatives and he embraced thermodynamics as well as elasticity (as his first book on theory of elastic solids (1884) indicates, in spite of its title; for there, he was concerned to retain a broad approach, to the extent of including intermolecular aspects of energy). He derived (1884) the variation of what became known as the complementary energy of an elastic solid but, unlike Crotti, did not develop it, though he referred to the energy method of Fränkel, as well as to that of Castigliano with application to frameworks. In the same work (like Müller-Breslau) he derived what is commonly known as the device of 'tension coefficients' for dealing with the equilibrium equations of space frameworks.

His paper on energy theory and statically-indeterminate systems (1886) is noteworthy (as is his article on Mayer's work on thermodynamics). Weyrauch was indeed an accomplished scientist generally, as well as an outstanding contributor to the further development of theory of structures, with particular reference to bridge engineering.

Notes

Wilson (1897) seems to have been among the first to use the concept of influence or flexibility coefficients in Britain, in his investigation of continuous beams. His notation is unique but somewhat tedious.

Weyrauch contributed significantly to theories of ultimate working strength of materials for design purposes (1880) and to a new theory of retaining walls (Erddruckes) in *Zeitschrift für Baukunde* (1878).

11

Secondary effects in structures

For the purpose of this chapter, secondary effects are understood to include dynamic stresses as well as those which arise from the nature of construction details, especially the rigidity of joints (connections) in triangulated trusses (bridge girders). The term secondary stress is usually associated with these latter, following the initiative of Professor Asimont of the Munich Polytechnikum in 1877. In that year a prize was offered by the Polytechnikum (as noted in Chapter 1) for a method of calculating those stresses in trusses (termed *Sekundarspannung* by Asimont, to distinguish them from the stresses due to the axial or primary forces in the bars, that is, *Hauptspannung*). Dynamic stresses, on the other hand, became the subject of research in the nineteenth century, due to the failure of a number of iron railway bridges, which were caused by the passage of trains.

Secondary stresses

According to Grimm (1908), Asimont formulated the problem of secondary stresses in rigidly-jointed trusses and suggested that, since the resultants no longer pass through the panel points, a solution might be afforded by 'Euler's equation of the elastic line'. In the event, the prize was awarded in 1879 to Professor Manderla and his solution was published soon afterwards (1880), although an approximate solution by Engesser in which chords were treated as continuous and web members as pin-jointed, appeared a year earlier (1879). Manderla's solution also contained the principles of analysing frameworks with rigid joints, whose bars resist loads, primarily by bending; and it included the relationships between terminal bending couples and the slopes and deflexions of bars.

Soon after Manderla's solution was published, the substance of a lecture in Berlin, by Winkler, on secondary stresses appeared (1881b) in which he remarked that for some years past he had been investigating the subject. Then Landsberg (1885) contributed a graphical solution based on an assumption which was similar to that of Engesser, whereby the joints of a truss provide continuity of the chords only. In view of the complexity of Manderla's method, analytical solutions were contributed by Müller-Breslau (1886a) and Mohr (1892), in order to ease the process of analysis; and W. Ritter (1890, vol. 2) introduced a graphical aid to the analytical process. Engesser's celebrated book (1892) on secondary stresses, provided a comprehensive treatment of stresses, other than primary stresses, in bridge trusses and it seemed to be adopted as the standard work of reference. But Winkler's masterpiece *Theorie der Brücken* (1881a, vol. 2) contains an extensive and impressive treatment of secondary stresses in bridge trusses.

The researches of Manderla, Winkler, Mohr, Müller-Breslau and Ritter, included the assumption that the rigidity of joints of trusses had a negligible effect on the deflexions of joints, due to imposed loads. (Indeed, since the rigidity of joints would be expected to reduce deflexions slightly, this assumption erred on the side of safety, it seemed.) The first step in calculating secondary stresses consisted, therefore, in finding the deflexions of joints by analytical or graphical means, on the basis of a pin-jointed framework. Then the bending of the rigidly-jointed bars was analysed for compatibility with the calculated deflexions, and the secondary stresses were evaluated. Manderla's investigation of the bending of the bars was very detailed and included the effect of change in bending stiffness, due to the axial or primary forces. When that effect is neglected, his analysis, like that of Winkler, revealed the equations:

$$
\left.
\begin{aligned}
\phi_1 &= \frac{l}{6EI}(2M_1 - M_2) \\
\phi_2 &= \frac{-l}{6EI}(M_1 - 2M_2)
\end{aligned}
\right\}
\qquad [(11.1)]
$$

(in Winkler's notation $\tau' = (2m' - m'')a/6EI$ and $\tau'' = -(2m'' - m')a/6EI$) for a bar of length l (or a) and flexural rigidity EI, due to terminal couples M_1 (or m') and M_2 (or m'') (Fig. 65) where ϕ_1 (or τ') and ϕ_2 (or τ'') are the changes in slope of the ends of the bar, due to those couples.

Müller-Breslau's attempt to simplify the analysis suffered from a surfeit of equations for all except the simplest framework. He neglected the effect of the primary forces in bars on their stiffness in bending, presumably on the frequently justifiable assumption that the bars would be of substantial section before secondary stresses merited investigation. Then, having determined the geometrical consequences for individual panels of deflexions

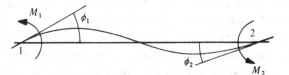

Fig. 65

of joints, due to primary stresses, he obtained relationships in terms of terminal bending moments of bars to satisfy the condition that each joint rotated as a whole without change of the angles between the connected bars. Having introduced the condition of zero resultant couple at each joint, sufficient equations emerged to enable all bending moments to be found, and thence the determination of the secondary stresses. Müller-Breslau further developed the application of his method to obtain influence lines for bending moments in the individual bars of a truss. Grimm (1908) gives a concise account of the method.

While differing from Müller-Breslau's method in conceptual detail, Ritter's method also concentrates on the terminal bending moments of bars as unknowns, but the equations to be solved simultaneously are arranged in a manner whereby a graphical analogy, using the funicular polygon, alleviates tedious computation.

Like Winkler but in contrast with some earlier investigators of secondary stresses, Mohr prudently chose the angles of rotation of joints of a truss, caused by secondary bending of bars, as unknowns, since they are related to the terminal bending moments of a bar by the inverse of equations (11.1). Thus, for any bar ij connecting joints i and j (Fig. 66):

$$\left.\begin{aligned} M_i &= \frac{2EI}{l_{ij}}(2\phi_i + \phi_j - 3\psi_{ij}) \\ M_j &= \frac{2EI}{l_{ij}}(\phi_i + 2\phi_j - 3\psi_{ij}) \end{aligned}\right\} \qquad\qquad [(11.2)]$$

where: M_i is the bending couple at the end of the bar at which it is connected to joint i; and M_j is the bending couple at the other end of the bar; ϕ_i and ϕ_j are the (small) rotations of joints i and j of the truss; and

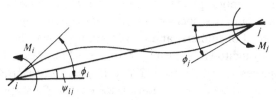

Fig. 66

ψ_{ij} is the small angular rotation of a straight line through those joints; EI and l_{ij} are the flexural rigidity and length of the bar ij. The quantities ψ_{ij} may be calculated with sufficient accuracy from the primary or axial strains of the bars, and this Mohr achieved by using the principle of virtual work. There then remain the equilibrium conditions, that for each joint:

$$\Sigma M_i = 0 \qquad\qquad\qquad [(11.3)]$$

since there is no resultant couple at a joint and these equations are clearly as numerous as the unknowns ϕ_i (representing joint rotations and changes of slope of the ends of bars which are rigidly connected together at individual joints). With the values of ψ_{ij} known, they are sufficient to complete the solution. Subsequently, the bending moments (and thence the secondary stresses in bars) are found by equations of the kind (11.2).

Mohr also considered means of alleviating the labour in the solving of simultaneous equations of the type (11.3), including aspects of symmetry, wherever possible, and successive approximations.

The slope-deflexion method for rigidly-jointed frameworks in general, was clearly foreshadowed by the solution of the problem of secondary stresses, especially by Manderla, Winkler and Mohr.

Dynamic effects

The study of the dynamics of structures was stimulated, it seems, mainly by the effect on iron railway bridges of the heavy loads moving at speed. But the failure of iron-chain bridges due to violent oscillation had also aroused the interest of scientific investigators who concentrated initially on the problem of the dynamic behaviour of an inelastic suspension chain (Rohrs, 1851; 1856). Then Tellkampf (1856), in his review of the practical theory of suspension bridges, includes an elementary approach to the problem of their oscillations, which is based on the theory of impact. Indeed, the problem is mentioned by Moseley (1843) within his account of the theory of resilience and impact, for which he acknowledges Poncelet's influence, especially with regard to the use of energy in the study of the dynamics of elementary elastic systems (such as an elastic rod subjected to a load suddenly applied to its extremity). Thus, Moseley refers to the fracture of the chains of suspension bridges due to oscillations caused by the 'measured tread of soldiers' and in a footnote (p. 496) he quotes the example of the Broughton Suspension Bridge, near Manchester. In the same footnote he quotes Navier (from his *Traité des ponts suspendus*, 1823) as showing that the duration of oscillations of the chains of a suspension bridge may extend to nearly six seconds, and he continues: 'there might easily, in such cases, arise that isochronism at each interval, or after any number of intervals, between the marching step of the troops and the oscillations of the bridge, whence would result a continually increasing elongation of the suspending chains'. The latter aspect refers to his elementary treatment (in the same chapter) of the phenomenon which later came to be known as 'fatigue of metal', due to repeated variations of load, being the ultimate reason for the fracture of suspension chains and consequent collapse of the bridges. The true complexity of the dynamic behaviour of suspension bridges was not, however, to be understood until the twentieth century, but safety was achieved (notably for the Niagara Bridge, 1855) by the adoption of systems of auxiliary stays and girders of relatively high stiffness. Incidentally, it is noteworthy that by 1857 Cubitt and Vignoles were advocating the stiffened suspension bridges as an economical and safe form for railway purposes. This they apparently did at the Dublin meeting of the British Association for the Advancement of Science (Todhunter, 1893), quoting the successful Niagara railway bridge and emphasising the saving in weight of material with regard to the economy afforded by this form of bridge construction.

Homersham Cox (1848) followed Moseley in applying Poncelet's treatment of resilience, in an article on the dynamical deflexion of railway bridges (presumably he was stimulated by the circumstances relating to the Royal Commission of 1847, noted in Chapter 1). Using the law of conservation of energy in the form of the principle of *vis viva*, with simplifying assumptions involving neglect of the kinetic energy associated with actual horizontal motion of the travelling load causing the deflexion, Cox deduces that the maximum statical deflexion which that load would produce cannot be more than doubled when motion takes place at any velocity. This conclusion is criticised by Stokes (1849) at some length, who nevertheless writes that Cox treats the subject in 'a very original and striking manner'.

Later (1849), Cox addresses himself to the problem of the deflexion of a uniform elastic beam when it is struck at its mid-point by a mass, in the form of a ball, moving horizontally in a direction normal to the axis of the beam, assuming *inter alia* that after the collision the ball becomes attached to the beam. For this purpose he derives the equivalent mass of the beam, with reference to its mid-point, by considering its kinetic energy in an assumed mode of dynamic deflexion. Then he uses the equation of conservation of momentum to determine the velocity of the mid-point after impact, in terms of the initial velocity of the ball, and finally he uses the equation of conservation of energy to calculate the resulting maximum elastic deflexion of the beam. According to Todhunter (1886) Cox's result is in general agreement with an experimental study of the problem by Eaton Hodgkinson. It is noted that Navier (1823) considered the problem of longitudinal impact of elastic bars, a problem which attracted the attention of his pupil, Saint-Venant (1867) who earlier (1857) had attacked the problem of transverse impact on an elastic beam (and its resulting vibration) in a more sophisticated manner than that adopted by Cox. (Saint-Venant's work on longitudinal shock and vibration was reported in Britain by Rankine in 1867 in a short article in *The Engineer*.)

It was the Royal Commission, appointed on 27 August 1847, 'for the purpose of inquiring into the conditions to be observed by engineers in the application of iron in structures exposed to violent concussion and vibration', following the fatal accident when an iron railway bridge at Chester collapsed (as noted in Chapter 1), that stimulated perhaps the most significant research of the century into dynamics of structures. The Rev. Robert Willis, Jacksonian Professor of Natural Philosophy at Cambridge and a member of the Royal Commission, undertook an extensive experimental investigation of the effect of travelling loads on metal structures, using facilities for large-scale experiments at the Royal Dockyard at Portsmouth. In this work he was assisted by Captain Henry James, R.E., F.R.S. and Lieutenant Douglas Galton, R.E. The results of their study are contained in appendix B of the 'Report of the commissioners' of 1849. At the same time, Willis attempted mathematical analysis and obtained a differential equation, for the solution of which he consulted Stokes whose paper (1849) on the subject contains an extensive analytical investigation of the whole problem, including, as noted above, comments on the work of Homersham Cox. It was assumed at first that the mass of the travelling load is large in comparison with that of the structure but, subsequently, Stokes examined the other extreme when the mass of the load is negligible and found that the effect is produced essentially by a constant force traversing the structure.

The Portsmouth experiments indicated that dynamical deflexion, of as much as three times the statical value, is likely at higher speeds of the load, but measurements on actual bridges did not appear to confirm that speed had such a marked effect. Willis accordingly investigated (at Cambridge) the discrepancy, using an extremely simple experiment, and he found that the effect when large deflexions are induced is due to forces associated with the trajectory of the load. Stokes provided an analytical solution in support of these findings but succeeded in obtaining only an approximate solution of what emerged as the more relevant problem of the moving load whose mass is much less than that of the structure. He showed that, then, the dynamical deflexion depends upon the ratio of the periodic time of the fundamental mode of vibration of the structure to the time taken for the passage of the load over the span.

Poirée (1854) gives details of experiments on arched ribs to determine deflexion, due to various effects, including impact and travelling loads such as railway trains. He obtains some measure of agreement with the results of the researches of Willis and Stokes.

Willis and Stokes were followed by Phillips (1855), Mallet (1860), Winkler (1860), Renaudot (1861) and Kopytowski (1865). Ultimately, Saint-Venant's treatment (1883) of the problem was published, which included his corrections to the work of those earlier investigators, including Willis and Stokes. But the essential value of the results of Willis and Stokes, in relation to the behaviour of iron railway bridges, remained unimpaired: otherwise, Saint-Venant's work alone represented a significant improvement in principle.

Mallet (1860) proposes a simple method of determining the deflexion of bridges, due to travelling loads, similar to that of Cox (1849), though he quotes Morin (1853). Also, Mallet applauds the work of Willis, Stokes and Phillips but regards that of the first as too sophisticated for practical men and does not appear to acknowledge that acceptance of their findings is at variance with his own theory.

Winkler's treatment of the problem of travelling loads, contained in his memoir of 1860, is inexact and reminiscent of Homersham Cox whose work, like that of Stokes and Phillips, seems to have been unknown to him.

Renaudot (1861) concerned himself with the effect of a rapidly travelling continuous load, a problem which was investigated by Kopytowski (1865) who used similar assumptions and an approach closely related to that of Phillips. Kopytowski also refers to Willis and Stokes, but his main objective is to extend Renaudot's work in detail. The value of his research is limited by the same errors as those made by Phillips and Renaudot. Indeed, all these researches are of somewhat trivial historical interest nowadays, with the exception, of course, of those of Stokes and Saint-Venant, which have an enduring quality. Todhunter (1893) reviews them all in some detail.

Toward the end of the century, the study of the dynamic aspect of structural behaviour was greatly enhanced by the work of Rayleigh and, especially, by his *Theory of sound* (1877). Indeed, it may be said that Rayleigh's work heralded the modern era of dynamics of elastic systems including, especially, engineering structures.

Appendix I
A note on C. L. M. H. Navier

The following is an abridged version of the author's free translation of Navier's Obituary Notice of 1837, by Prony, which is included in the 1864 edition of Navier's *Leçons*, edited and with additional notes by Saint-Venant.

Louis-Marie-Henri Navier, officier de la Légion d'Honneur, member of l'Institut Royale de France and Divisional Inspector of le Corps Royale des Ponts et Chaussées, was born in Dijon on 15 February 1785. His father was a lawyer of distinction and died at an early age as a result of the excesses of the revolutionaries.

Navier, orphaned at fourteen years of age, had the good fortune to find a second father in an uncle, numbered with reason among the notables of le Corps des Ponts et Chaussées, M. Gauthey, who, having been an engineer for the region of Bourgogne, became Inspector General of bridges and highways following the departmental organisation of France; he died 14 July 1807, after having designed and accomplished works of the greatest importance including the remarkable constructions mentioned later in this notice.

Navier's education, supervised by Gauthey, had (not surprisingly) that emphasis on scientific culture which would be familiar to an engineer; the progress of the young pupil was such that in 1802 he presented himself for examination for admission to l'Ecole Polytechnique and was among the highest in order of merit; after a brilliant record there, he entered l'Ecole des Ponts et Chaussées in 1804, and in 1808 obtained the qualification of ordinary engineer.

Following Navier's courses at l'Ecole Polytechnique and l'Ecole des Ponts et Chaussées, Gauthey devoted what little amount of leisure time his last four or five years allowed to interesting his nephew in those aspects of his own work which would be useful to the young student. The result of that association was the attitude and outstanding facility acquired by Navier for dealing with problems wherein theory and practice are intertwined, a facility which had tremendous influence for good on his later work. With such a background he found himself able to make the best use of his ability, in an epoch of life in which the majority of those who had inclinations to the sciences were denied facilities for instruction. His first major publications were acclaimed and his scientific career began. At the time of his death,

Gauthey had nearly completed his accounts of bridges and canals and these were eagerly awaited: Navier, who did not inherit his uncle's fortune, made great sacrifices to obtain the rights of those manuscripts and complete them for publication.

The *Traité des ponts* appeared in 1813 [Author's Note: here Prony appears to have overlooked the fact that the first volume appeared in 1809] and consisted of two large volumes. (The first volume was reprinted in 1832 with substantial annotations and additions.) This treatise, the most complete of all to have been published on the same subject was, above all, noteworthy for the improvement which it afforded in the actual level of knowledge, and for Navier's addition of a large number of notes which, in themselves, provided a very instructive separate work.

The first volume contained historical details and descriptions of bridges (both ancient and modern) which are, in themselves, of much interest, not only to engineers, but also to those who, while not possessing scientific knowledge about them, are simply interested in the history of their art. Navier provided, at the beginning of this volume, a biographical note about his uncle.

Though this treatise on bridges first appeared in 1813, it would probably have appeared earlier, if Navier had not been interrupted by a mission to Rome on the instructions of M. le comte Mole (Director General of bridges and highways) to undertake important projects: namely the reconstruction of the bridge of Horatius Cocles, numerous quays throughout Rome and ways of protecting the holy city from encroachment of the Tiber.

The writer had the satisfaction of being in Rome when Navier was working on those important projects which are now in abeyance due to well-known political events.

A third volume of the works, left in manuscript by Gauthey, and published by Navier, deals with navigable channels. This volume, date 1816, gives a collection of seven memoirs containing (independent of general matters concerning construction of canals) very detailed notes on the main canals of France, as constructed rather than planned, and particularly the Canal du Charolais, which has subsequently taken the characteristic name of the 'Central Canal' and which was described by Gauthey who also supervised its construction.

Navier, in publishing his uncle's manuscripts, had fulfilled a heart's desire. Love of science alone also made him undertake the republication of two other commendable works, by Bélidor, which had been rewritten and brought up to date during the first part of the last century, namely *La science des ingénieurs dans la conduite des traveaux des fortifications d'architecture civile* (1729) and *Architecture hydraulique* (1737, vol. 1). Navier's edition (1813) of Bélidor's book of 1729 is enriched with numerous notes and additions among which are those concerning earth pressures, the form and dimensions of retaining walls, and the theory of arches. Volume 1 of Bélidor's later work (1837) is a treatise on mechanics in which the theories of equilibrium and motion of solid bodies and fluids are presented with their applications in engineering practice.

In addition to scientific work, Navier was concerned with the construction of three bridges over the Seine: the bridges of Choisy, Asnières and Argenteuil, which consisted of circular stone arches. The interest of the Government in suspension bridges and railways caused the Directorate General of bridges and highways to give every facility to enable French engineers to complete their studies by becoming familiar with these new kinds of construction; and Navier, having been sent to England and Scotland several times, was charged with the task of providing the information and documentation for that purpose.

The memoir prepared by Navier on suspension bridges, which he published in 1823 (after his first two visits), has been judged a very complete and original account of the subject. Its formal recognition by the Academy of Sciences followed a report on it by Messieurs Prony, Fourier, Fresnel, Molard and Charles Dupin; the author, by his recording and evaluation of existing knowledge of suspension bridges, being credited with originality and depth of understanding of theoretical and practical mechanics.

From this time, Navier was consulted widely regarding the engineering of structures in France and abroad.

Soon after, Navier was named a member of l'Académie Royale des Sciences and this reward, at so early an age, surpassed his greatest hopes. Unfortunately, it was the destiny of this distinguished engineer to provide a fresh example of the hazards which, from time to time, upset the lives of the most talented. Navier was at the peak of his career when he had to suffer an unjust and cruel injury to his reputation as an engineer: it will be obvious that I am referring to the Pont des Invalides.

This bridge, of 155 m span, was nearly equal in length to that of Telford across the Menai Straits.

For the purpose of understanding the merit of the theoretical and experimental research which had been done with regard to finding the forces on the various elements of the Pont des Invalides, it suffices to recall that the chains were to withstand forces of four-and-a-half times greater than those of the bridge built in 1820 over the River Tweed by Captain Brown, a bridge which was then the largest available for study.

Paris had looked forward to the monumental spectacle of this huge curve with four Egyptian columns; engineers themselves were eager to inspect the novel structural features and details of this great structure, devised by Navier for a load of 67000 kg (the weight of nearly one thousand men).

The bridge was practically finished when a slight movement appeared in the chain anchorage pits: this movement was aggravated on the Champs-Elysées side by the fracture, during the night of 6–7 September 1826, of one of the city's water conduits, which had a diameter of 0.32 m. The resulting torrent of water (under a head of 30 m) from the Chaillot reservoirs formed a pool alongside incomplete excavations and led to infiltration of earthworks which were already finished.

It was necessary from that time to put the structure on scaffolding; the season was late, there was dread that the temporary works would not be dismantled before the extreme high tides and winter frosts; moreover, a paltry matter of money delayed the beginning of the emergency work which was necessary to consolidate the anchorages of the suspension chains, and it was decided to adjourn repairs till the following season and to dismantle the deck and the chains.

Sadly, it must be recorded that national pride in progress was severely deprived because the work of the Pont des Invalides was never made good. On the contrary, work that had been completed was demolished and no trace was left of this great and beautiful enterprise.

The demolition of the Pont des Invalides seems to have been due to three factors:

1. opposition by the City of Paris to the construction of a road crossing the Champs-Elysées to reach the bridge;

2. disagreement between the company and administration, concerning payment for the work of reinforcing the foundations;

3. the transaction, over and above these difficulties, by which the company became obliged to give up the Pont des Invalides in order to retain concessions relating to the construction of three other bridges over the Seine.

These details about the Pont des Invalides were published at the time by order of the administration (*Monitem*, 29 February 1828) and reproduced with very interesting elaborations by Navier, in the 1830 edition of his memoir on suspension bridges, under the heading 'Notice sur le Pont des Invalides'.

On his first tour of Britain, Navier gave attention to the excellent roads in that country and reported to the administration in 1822 regarding the Macadam process, as published in *Annales des Ponts et Chaussées* (1832, 1st semester). This study was concerned as much with the remarkable consequences of the method of construction as with the details of the method, namely:

1. improved drainage.

2. foundations to eliminate irregularities of the ground.

Later, Navier's collaboration with leading British engineers and his multifarious researches on matters arising from his report on roads, including the usage and the legislation governing them, led to the publication of his memoir on road traffic. His opinions were controversial but the outcome of public opinion and discussions increased his standing among engineers.

A new tribute was to be paid to Navier's special studies concerning such matters: a commission had been formed by the Director General of bridges and highways to produce regulations for road traffic on the basis of that knowledge and Navier was chosen by the administration to serve on the commission and was unanimously elected by his colleagues to be the secretary and reporter of its work.

Navier was also among several engineers sent to Britain to study railways, and the remarkable report of those engineers, in *Annales des Ponts et Chaussées*, was due to his observations and combination of practical and scientific knowledge: the report was used as the basis for the construction of the main railway line from Paris to Strasbourg.

Navier was awarded la Croix de la Légion d'Honneur in recognition of his indefatigable services.

As well as leading in practical engineering projects, Navier excelled also as a professor and as an academician.

Appointed (in 1819) supervisory professor of applied mechanics at l'Ecole Royale des Ponts et Chaussées, and titular professor in 1831, he discharged his duties to the great advantage of both his pupils and science. His lectures (lithographed and distributed to the students who had attended his courses and, in general, widely circulated among engineers) afforded the Corps that instruction which was increasingly necessary for the greatest advantage to be taken of scientific principles in practice.

A vacancy for a professor of analysis and mechanics at l'Ecole Royale Polytechnique occurred at the end of 1831, Navier was appointed and filled the office for the last six years of his life. He is renowned in this famous school for his methods and the clarity of his work and he performed a great service in lithographing his lectures, the whole of which formed four volumes (two for the first year and two for the second year), one volume of each pair containing analysis and the other, mechanics.

Navier was distinguished within l'Académie Royale des Sciences, to which he was elected 26 January 1824, for his outstanding services to that learned body, including the numerous reports on works that he submitted to it.

On 23 August 1836 an illness carried Navier away unexpectedly from the sciences and his numerous friends and colleagues; I say his numerous friends because he could count among them, all those with whom he associated frequently. Although by nature a little stolid, he was none the less given to sincere and durable friendships, and with this disposition were other qualities with which he was endowed in high degree; the details of his private life are as honourable as his public life and commensurate with his talent.

A cortège of engineers and of pupils accompanied Navier's funeral procession: three engineers, Messieurs Emmery, Coriolis and Rancourt gave addresses at the time of burial and their brief but meaningful speeches were heard with close attention.

Appendix II
A note on Carl Culmann

Carl Culmann was born 10 July 1821 in Bergzabern, Rheinpfalz, and died in Zurich, 9 December 1881.

After completing his studies in Karlsruhe he 'worked on railway construction in mountainous country and later (1848) was transferred to the office of the Royal Railways Commission in Munich.

In the summer of 1849 the Railways Commission sent him on a two-year study tour of the British Isles and the U.S.A. The period of this tour coincided with the completion of the wrought iron Britannia (tubular) Bridge by Robert Stephenson and with the end of a phase of intensive development of wooden bridge construction in the U.S.A. The substance of Culmann's report of the tour was published in *Allgemeine Bauzeitung* in 1851 under the title 'A description of the latest advances in bridge, railway and river-boat construction in England and the United States of North America'. It aroused great interest and established Culmann's reputation as a young engineer with outstanding qualities of perception. Indeed, it seems to have been a material factor in his leaving the railway industry in 1855 to teach at the newly established Federal Polytechnic Institute at Zurich, where he believed he would have greater opportunities for combining theory and practice of engineering.

Culmann clearly recognised the urgent need to develop Navier's methods for application to the design of railway bridges and his report emphasised methods of calculating the forces in the new bridge forms to enable them to be exploited with confidence in their safety.

Culmann was not alone in recognising the need for precise theory or in attempting to revolutionise the teaching of construction statics of his time for, in the same year (1851) that his report was published, Schwedler published a report on his own investigation, with essentially the same conclusion. If, later, Culmann's work is to be regarded as more significant, this is to be attributed more to the great regard in which his later achievements were held than to any superiority of his findings over those of Schwedler.

Timoshenko (1950) has shown that, before Schwedler and Culmann, structures had been accurately analysed by Jourawski, in Russia, and even before him by Whipple, in the U.S.A., who had published a book entitled *An essay on bridge-building* (1847), which contained such structural analysis. It is astonishing that Culmann – to judge from his report – did not know of this book. Perhaps he paid little heed to earlier analysis (after the fashion for failure to mention earlier work) because he felt himself capable of carrying out the investigations alone and independently.

The almost simultaneous development of a theory of structures by four different engineers, whose individual independence is scarcely to be doubted, probably had its origin, not only in a strong demand from the world of engineering of that time, but also in the existence of those elements of the theory, which invited development.

But Culmann was unique in his insight into the power of graphical techniques of analysis. French engineers, like Poncelet and Cousinery, had indeed already sought graphical solutions, but they were merely either substituting drawn constructions for certain computational steps, or were translating former methods into the language of drawing.

Culmann's goal was more revolutionary: he sought to derive geometrically, the relationships occurring in the theory of structures. His employment of the newer geometry, the 'geometry of situation', afforded him insight into important structural relationships and led him, by clearly arranged and vivid ways, into the graphical language of the engineer.

Of even greater significance than its practical application, graphical statics appears to have influenced the development of structural analysis generally. In 1875 Culmann had, in a second edition of his book (1866), published (in a much extended form) the general foundations of his teachings; he was not to be allowed, however, to complete his intended second volume which was to include applications. His pupil and successor, at Zurich, Wilhelm Ritter, continued the work instead (1888–1907).

In the foreword to his second edition, Culmann is enthusiastic about the advances in graphical statics since the appearance of the first edition, for he is quoted by Stussi (1951) as saying with regard to the reception of his theories:

In Italy they found indisputably the most favourable ground. Cremona has introduced them at the Milan Polytechnic and, indeed, in an advanced form: he does not view them simply as a practical aid just to avoid a few calculations in certain cases, but as the completion of the geometric-statics-education of the young engineer.

In a short comparison of the reciprocal figures of Maxwell with those of Cremona, he writes: 'And this introduction of the null-system is the work of Cremona and not Maxwell; also from Cremona stem all the applications. More than anywhere else the gulf between the theoretician and the practitioner yawns in England.'

Culmann appears equally pleased with the dissemination of his methods in France, for he mentions most favourably the textbook by Levy (1874). He rejects, on the other hand, the corresponding efforts in Germany – to the point where he could be argumentative and unfriendly about them. He was especially critical of Bauschinger of Munich, whose book (1871) purported to be based on Culmann's work. In the event, however, the later intense development of graphical methods of analysis took place in Germany and included the elegant and fruitful use of the elastic line for structures by Otto Mohr.

One of the most noteworthy engineering achievements of the Culmann school was, incidentally, the Eiffel Tower in Paris, built for the Paris Exhibition of 1889 and for which the design and calculations were carried out by Culmann's pupil Maurice Koechlin (1889a).

The calculations for this structure (which was, for its time, unbelievably bold) are classical applications of the Culmann graphical statics. They are, incidentally, described in full detail in a monumental luxury volume (1890) by the constructors. Maurice Koechlin has also rendered valuable service to the dissemination of Culmann's teaching in a textbook (1889a).

Appendix III
A note on John Robison

The state of knowledge of applied mechanics in Britain at the beginning
of the nineteenth century is probably reflected in David Brewster's
Robison's Mechanical Philosophy (1822) which is based mainly upon
articles published by John Robison, professor of natural philosophy at
Edinburgh University, in the fourth edition of the *Encyclopaedia Britannica*
(1797). The first of the four volumes includes chapters on strength of
materials, carpentry, roofs, construction of arches and construction of
centres for bridges. With regard to strength of materials Robison refers
(in general terms) to the experiments of Couplet, Pitot, De La Hire and
Duhamel in relation to cohesion. He also refers specifically to elasticity and
ductility and mentions plastic substance and properties. Robison is much
concerned with cohesion in terms of attraction between particles, referring
to the theories of Newton and Boscovich. Then he suggests that 'connecting
forces are proportional to the distances of the particles from their
quiescent, neutral or inactive positions'. This 'seems to have been first
reviewed as a law of nature by the penetrating eye of Dr Robert Hooke'.
Robison quotes what he describes as Hooke's cipher, *ceiiinosssttu*, for the
law of elasticity (*ut tensio sic vis*) which bears his name and he records
Hooke's anticipation – and rejection – of the facts used by John Bernoulli
in support of Leibnitz's doctrine of *vires vivae*. Then, Robison considers
James Bernoulli's observation of the relationship between strain and
curvature of a bar (the *elastic curve*) and applauds the elegance of his
mathematical treatment of the problem as published in 1694 and 1695
(*Acta Lipsiae*). He goes on to credit Daniel Bernoulli, nephew of James,
with an elegant contribution to the same problem.

Among the other names mentioned by Robison concerning strength
of materials are Muschenbroek, Euler, Parent, Gauthey, Mariotte and

Varignon. In emphasising Hooke's priority (*Proceedings of the Royal Society*, 1661) with regard to the linearity of the elasticity of common materials, he asserts:

Mariotte indeed was the first who expressly used it for determining the strength of beams: this he did about 1679, correcting the simple theory of Galileo. Leibnitz indeed, in his dissertation *de Resistentia Solidorum*, in the *Acta Eruditorum* (1684) introduces this consideration, and wishes to be regarded as the discoverer; and he is always acknowledged as such by the Bernoulli's and others who adhered to his peculiar doctrines! But Mariotte had published the doctrine in the most express terms long before; and Bülfinger, in the *Commentarri Academiae Scientiarum Imperialis Petropolitanae*, 1729, vindicates his claim.

In drawing attention to errors in Bülfinger's dissertation of 1729, in respect of the fibres in beams, Robison says: 'We recommend to the reader's perusal the very minute discussions in the memoirs of the Academy of Paris for 1702 by Varignon, the memoirs for 1708 by Parent, and particularly Coulomb (1776).'

Robison discusses Euler's theory of the strength of columns, ascribing priority to him; but he does not enter into mathematical detail, nor does he associate the problem with *tottering* equilibrium, a concept which he introduces in his chapter on arches. He says, however:

Fortunately the force requisite for crippling a beam is prodigious, and a very small lateral support is sufficient to prevent that bending which puts the beam in imminent danger. A judicious engineer will always employ transverse bridles...to stay the middle of long beams, which are employed as pillars, struts or truss beams, and are exposed, by their position, to enormous pressures in the directions of their lengths.

He also quotes the practice of Perronet, and Bélidor's *Science des Ingenieurs* (1729) in this respect.

Robison's theory of bending of beams contains the error of his times regarding the position of the neutral axis. But it is noteworthy that he argues correctly that the strength of an encastré beam is twice that of a simply supported beam and is at some pains to describe the conditions necessary for provision of a truly encastré state (including full continuity, for example).

In addition to strength of materials (especially bending and rupture of beams) in which he mentions the 'power' of strain with regard to the work of straining forces, Robison deals with the elementary statics of (timber) framework and masonry arches (crediting Hooke with the concept of the inverted hanging chain for the shape of an arch). He illustrates a variety of timber frameworks, especially roof trusses, including one consisting of three equilateral triangles, which was precisely the same as that which came to be known as the Warren girder. The other timber framework relates to centring for construction of masonry arches. The work as a whole is primarily descriptive: there is some elementary analysis concerning equilibrium of systems of bars and straining of beams as well as that relating to arches but (although he refers to original researches, for example of Euler, concerning columns) he does not reproduce the relevant analytical matter.

Bibliography

Anon. (1860). Suspension girder bridges for railway traffic, *Civil Engineer and Architect's Journal*, 23, 317–19, 352–6.

Anon. (1862). The statics of bridges, *Civil Engineer and Architect's Journal*, 25, 246–9.

Am Ende, M. (1898). Suspension bridges with stiffening girders, *Minutes of the Proceedings of the Institution of Civil Engineers*, 137, 306–42.

Asimont, G. (1880). Hauptspannung und Sekundarspannung, *Zeitschrift für Baukunde*, 33, 116.

Baker, T. (1851). *The principles and practice of statics*. London.

Balet, J. W. (1908). *Analysis of elastic arches*. New York.

Barlow, W. H. (1846). On the existence (practically) of the line of equal horizontal thrust in arches and the mode of determining it by geometrical construction, *Minutes of the Proceedings of the Institution of Civil Engineers*, 5, 162–82.

Barlow, P. W. (1858). Combining girders and suspension chains, *Journal of the Franklin Institute*, 65, 301–9, 361–5.

Barlow, P. W. (1860). On the mechanical effect of combining girders and suspension chains and the application of the system for practical purposes, *Civil Engineer and Architect's Journal*, 23, 225–30.

Bauschinger, J. (1871). *Elemente der graphischen Statik*. Munich.

Belanger, J. B. (1858). *Théorie de la résistance et de la flexion plane des solides*, 2nd edn, 1862. Paris.

Bélidor, B. F. (1729). *La science des ingénieurs dans la conduit des traveaux des fortifications d'architecture civile*, 2nd edn, 1813, edited and with additional notes by C. L. M. H. Navier. Paris.

Bélidor, B. F. (1737). *Architecture hydraulique*, 2nd edn, 1819, edited and with additional notes by C. L. M. H. Navier. Paris.

Bell, W. (1871). On the stresses of rigid arches, continuous beams and curved structures, *Minutes of the Proceedings of the Institution of Civil Engineers*, 33, 58–165.

Bernoulli, D. (1751). Excerpta ex literis a Daniele Bernoulli ad Leonh Euler, *Commentarii Academiae Scientiarum Imperialis Petropolitanae*.

Bertelli, F. (1850). Ricerche sperimentali circa la pressione dei corpi solidi ne' casi en cui la misura di essa, secondo le analoghe teoria meccaniche di pressioni e la elasticità de' corpi medesimi, *Memoria Posthuma. Memorie dell'Accademia delle Scienze di Bologna*, 1, 433–61.

Bertrand, J. L. F. (1869). Abstract of letter to Menabrea, *Atti della Reale Accademia delle Scienze di Torino*, 5, 702.

Betti, E. (1872). Teorema generale intorno alle deformazioni che fanno equilibro a forze che agiscono soltanto alle superficie, *Il Nuovo Cimento*, ser. 2, **7–8**, 87–97.

Betti, E. (1913). *Opere mathematiche*, vol. 2. Hoepli, Milan.

Bow, R. H. (1851). *A treatise on bracing and its application to bridges.* Edinburgh.

Bow, R. H. (1855). Iron railway bridge designs of T. Bouch, *Civil Engineer and Architect's Journal*, **18**, 236.

Bow, R. H. (1873). *Economics of construction in relation to framed structures.* London: Spon.

Bresse, J. A. C. (1854). *Recherches analytiques sur la flexion et la résistance des pièces courbes.* Paris.

Bresse, J. A. C. (1859). *Cours de mécanique appliquée*, part 2, 1865; 2nd edn, 1866; 3rd edn, 1880. Paris.

Brewster, D. (1822). *Robinson's mechanical philosophy.* London.

Castigliano, C. A. P. (1873). *Intorno ai sistemi elastici.* Turin.

Castigliano, C. A. P. (1875). Intorno all'equilibrio dei sistemi elastici, *Atti della Reale Accademia delle Scienze di Torino*, **10**, 380–422; **11**, 127–86.

Castigliano, C. A. P. (1879). *Théorie de l'équilibre des systèmes élastiques et ses applications.* Turin: Negro. (English translation by E. S. Andrews, 1919. London: Scott Greenwood.)

Chalmers, J. B. (1881). *Graphical determination of forces in engineering structures.* London: Macmillan.

Charlton, T. M. (1960). Contributions of Navier and Clebsch to the theory of statically-indeterminate frames, *The Engineer*, **210**, 712–13.

Charlton, T. M. (1962). Some early work on energy methods in theory of structures, *Nature*, **196**, 734–6.

Charlton, T. M. (1963). Maxwell–Michell theory of minimum weight of structures, *Nature*, **200**, 251–2.

Charlton, T. M. (1971). Maxwell, Jenkin and Cotterill and the theory of statically-indeterminate structures, *Notes and Records of the Royal Society of London*, **26**, 233–46.

Charlton, T. M. (1976a). Contributions to the science of bridge-building in the nineteenth century by Henry Moseley, Hon. Ll.D., F.R.S. and William Pole, D. Mus., F.R.S., *Notes and Records of the Royal Society of London*, **30**, 169–79.

Charlton, T. M. (1976b). Theoretical work. In *The works of Isambard Kingdom Brunel*, Alfred Pugsley (ed.). London: The Institution of Civil Engineers and The University of Bristol. (2nd edn, 1980. Cambridge University Press.)

Charlton, T. M. (1978). A note on the contributions of Cotterill, Castigliano, Crotti and Engesser to an energy principle of structures, *International Journal of Mechanical Sciences*, **20**, 659–64.

Chasles, M. (1837). *Aperçue historique sur l'origine et le développement des méthodes en géométrie.* Paris.

Chasles, M. (1852). *Traité de géométrie supérieure.* Paris.

Clapeyron, B. P. E. (1857). Calcul d'une poutre élastique reposant librement sur des appuis inégalement espacés, *Comptes Rendus hebdomadaires des Séances de l'Académie des Sciences*, Paris, **45**, 1076–80; **46**, 208–12.

Clapeyron, B. P. E. (1858). Mémoires sur le travail des forces élastiques dans un corps solide élastique déformé par l'action des forces extérieures, *Comptes Rendus hebdomadaires des Séances de l'Académie des Sciences*, Paris, **46**, 208.

Clark, E. (1850). *Britannia and Conway tubular bridges.* London: Longman.

Clebsch, A. (1862). *Theorie der Elastizitat fester Körper.* Leipzig: Teubner. (French translation by B. de Saint-Venant & A. A. Flamant, 1883, *Théorie de l'élasticité des corps solides*, with additional notes by B. de Saint-Venant. Paris: Dunod.)

Coignet, E. & Tedesco, N. de (1894). *Du calcul des ouvrages en ciment avec ossature métallique.* Paris.

Collignon, E. (1869). *Cours de mécanique.* Paris.

Coriolis, G. (1844). *Traité de la mécanique des corps solides,* 2nd edn. Paris.

Cotterill, J. H. (1865*a*). On an extension of the dynamical principle of least action, *London, Edinburgh and Dublin Philosophical Magazine,* ser. 4, **29**, 299.

Cotterill, J. H. (1865*b*). On the equilibrium of arched ribs of uniform section, *London, Edinburgh and Dublin Philosophical Magazine,* ser. 4, **29**, 380.

Cotterill, J. H. (1865*c*). Further application of the principle of least action, *London, Edinburgh and Dublin Philosophical Magazine,* ser. 4, **29**, 430.

Cotterill, J. H. (1869). On the graphic construction of bending moments, *Engineering,* 7, 32.

Cotterill, J. H. (1884). *Applied Mechanics,* edns, 1890, 1892, 1895, 1900. London: Macmillan.

Coulomb, C. A. (1776). Essai sur une application des règles de *maximis & minimis* à quelques problèmes de statique, relatifs à l'architecture, *Mémoires de Mathématique et de Physique, présentés à l'Académie Royale des Sciences par divers Savans,* 7, 343–82 (written, 1773).

Cox, H. (1848). The dynamical deflexion and strain of railway girders, *Civil Engineer and Architect's Journal,* 11, 258–64.

Cox, H. (1849). On impact on elastic beams, *Cambridge Philosophical Transactions,* 9, part 1, 73–8.

Cremona, L. (1872). *Le figure reciproche nella statica grafica.* Milan. (English translation by T. Hudson Beare, 1890. Oxford: Clarendon Press.)

Crotti, F. (1888). *La teoria dell elasticita* (based partly on a submission to the Collegio Ingegneri di Milano, 1883). Milan: Hoepli.

Cubitt, J. (1852). A description of the Newark Dyke bridge on the Great Northern Railway, *Minutes of the Proceedings of the Institution of Civil Engineers,* 12, 601–12.

Culmann, C. (1851). A description of the latest advances in bridge, railway and boat construction in England and the U.S.A., *Allgemeine Bauzeitung,* 16, 69–129.

Culmann, C. (1866). *Die graphische Statik,* 2nd edn, 1875. Zurich.

De Maré, E. (1954). *The bridges of Britain.* London: Batsford.

Dempsey, G. D. (1850). *Iron roofs.* London.

Dempsey, G. D. (1864). *Tubular and other iron girder bridges particularly describing the Britannia and Conway tubular bridges.* London: Virtue.

Dorna, A. (1857). Memoria sulle pressioni sopportate dai punti d'appoggio di un sistema equilibrato ed in istato prossimo al moto, *Memorie dell'Accademia delle Scienze di Torino,* ser. 2, **18**, 281–318.

Doyne, W. T. (1851). Description of a wrought iron lattice bridge constructed over the line of the Rugby and Leamington railway, *Minutes of the Proceedings of the Institution of Civil Engineers,* 9, 353–9.

Doyne, W. T. & Blood, W. B. (1851). An investigation of the strains upon the diagonals of lattice beams with the resulting formulae, *Minutes of the Proceedings of the Institution of Civil Engineers,* 11, 1–14.

Du Bois, A. J. (1875). *Elements of graphical statics,* 2nd edn, 1877. New York: Wiley.

Du Bois, A. J. (1882). A new theory of the suspension system with stiffening truss, *Journal of the Franklin Institute,* 113, 117–33, 163–73, 241–55.

Dugas, R. (1955). *A history of mechanics.* Neuchâtel: Griffon.

Eiffel, G. (1890). La tour de trois cents metres, 2 vols. Paris.

Emmerson, G. S. (1972). *Engineering education: a social history.* Newton Abbot: David & Charles.

Engesser, F. (1879). Ueber die Durchbiegung von Fachwerkträgern und die hierbei auftretenden zusätzlichen Spannungen, *Zeitschrift für Baukunde,* 29, 94–105.

Engesser, F. (1880). *Theorie und Berechnung der Bogenfachwerkträger ohne Scheitelgelenk.* Berlin.

Engesser, F. (1889). Uber statisch unbestimmte Träger bei beliebigen Formanderungsgesetze und über den Satz von der kleinsten Erganzungsarbeit, *Zeitschrift des Architekten- und Ingenieur-Vereins zu Hannover*, **35**, col. 733.

Engesser, F. (1892). *Die Zusatzkräfte und Nebenspannungen eisemer Fachwerkbrücken.* Berlin.

Euler, L. (1744). *Methodus inveniendi lineas curvas maximi minimive proprietate gandentes, sive solutio problematis isoperimetrici latissimo sensu accepti.* Lausanne and Geneva.

Euler, L. (1751). Excerpta ex literis a Daniele Bernoulli ad Leonh Euler. *Commentarii Academiae Scientiarum Imperialis Petropolitanae*, **13**, 8.

Euler, L. (1774). De pressione ponderis in planum cui incumbit. *Novi Commentarii Academiae Scientiarum Imperialis Petropolitanae*, **18**, 289–329.

Fairbairn, W. (1850). On tubular girder bridges, *Minutes of the Proceedings of the Institution of Civil Engineers*, **9**, 233–41. Discussion, 242–87.

Favaro, E. (1879). *Leçons de statique graphique.* Paris.

Fidler, T. Claxton. (1878). On suspension bridges and arches and particularly a new form of suspension bridge, *Transactions of the Royal Scottish Society of Arts*, **9**, 185–99 (read, 1874 and published also in *Engineering*, **19** (1875), 372).

Fidler, T. Claxton. (1883). Continuous girder bridges, *Minutes of the Proceedings of the Institution of Civil Engineers*, **74**, 196.

Föppl, A. (1880). *Theorie des Fachwerks.* Leipzig.

Föppl, A. (1892). *Das Fachwerk in Raume.* Leipzig.

Fränkel, W. (1867). Berechnung eisener Bogenbrücken, *Der Civilingenieur*, **13**, 75.

Fränkel, W. (1875a). Anwendung der Theorie des augenblicklicken Drehpunktes auf die Bestimmung der Formveränderung von Fachwerk U.S.W., *Der Civilingenieur*, **21**, 121ff.

Fränkel, W. (1875b). Anwendung der Theorie des augenblicklichen Drehpunktes auf Bogenträger: Theorie des Bogenfachwerks mit zwei Gelenken, *Der Civilingenieur*, **21**, 585ff.

Fränkel, W. (1876a). Theorie des ein fachen Sprengwerkes, *Der Civilingenieur*, **22**, 22ff.

Fränkel, W. (1876b). Ueber die ungünstigste Einstellung eines Systems von Einzellasten auf Fachwerkträgern mit Hilfe von Influenzkurven, *Der Civilingenieur*, **22**, 218, 441.

Fränkel, W. (1882). Das Prinzip der kleinsten Arbeit der inneren Kräfte elastischer Systeme und seine Anwendung auf die Lösung baustatischer Aufgaben, *Zeitschrift des Architekten- und Ingenieur-Vereins zu Hannover*, **28**, col. 63–76.

Fuller, G. (1874). Curve of equilibrium for a rigid arch under vertical forces, *Minutes of the Proceedings of the Institution of Civil Engineers*, **40**, 143–9.

Gauss, C. F. (1829). Über ein neues Grundgesetz der Mechanik, *Crelles Journal für die reine und angewandte Mathematik*, **4**, 233.

Gauthey, E. M. (1809), *Traité de la construction des ponts*, vol. 1; vol. 2, 1813; vol. 3, 1816; 2nd edn of vol. 1, 1832; edited by C. L. M. H. Navier.

Gebbia, E. (1891). Una questione di priorita su alcune contribuzioni alla teoria dei sistemi articolati, *Il Politechnico*, **39**, 778–82.

Godard, M. T. (1894). Récherches sur le calcul de la résistance des tabliers des ponts suspendus, *Annales des Ponts et Chaussées*, ser. 7, **8**, 105–89 (also in *Minutes of the Proceedings of the Institution of Civil Engineers*, **137** (1895), 417–19).

Green, G. (1839). On the laws of the reflection and refraction of light at the common surface of two non-crystallised media, *Cambridge Philosophical Transactions*, **7**, 1–24 (read, 1837).

Gregory, D. (1697). Catenaria, *Philosophical Transactions of the Royal Society of London*, **231**, 637.

Gregory, O. (1806, 1815). *A treatise on mechanics, theoretical, practical and descriptive.* London: Craddock & Joy.

Gregory, O. (1825). *Mathematics for practical men.* London: Craddock and Joy.

Grimm, C. R. (1908). *Secondary stresses in bridge trusses.* New York: Wiley.

Hann, W. (1843). Theory of bridges. In *The theory, practice and architecture of bridges of stone, iron, timber and wire with examples on the principle of suspension,* J. Weale (ed.). London: Architectural Library.

Helmoltz, H. (1847). *Ueber die Erhaltung der Kraft.* Berlin.

Hemans, G. W. (1844). Description of a wrought iron lattice bridge, *Minutes of the Proceedings of the Institution of Civil Engineers,* **3,** 63–5.

Henneberg, L. (1886). *Statik der starren Systeme.* Darmstadt.

Heppel, J. M. (1860). On a method of computing the strains and deflections of continuous beams under various conditions of load, *Minutes of the Proceedings of the Institution of Civil Engineers,* **19,** 625–43.

Heppel, J. M. (1870). On the theory of continuous beams (with comment by W. J. M. Rankine), *Proceedings of the Royal Society of London,* **19,** 56–71.

Hetenyi, M. (1946). *Beams on elastic foundations,* Ann Arbor, Michigan.

Heyman, J. (1966). The stone skeleton, *International Journal of Solids and Structures,* **2,** 249–79.

Heyman, J. (1969). The safety of masonry arches, *International Journal of Mechanical Sciences,* **11,** 363–85.

Heyman, J. (1972). *Coulomb's memoir on statics: an essay in the history of civil engineering.* Cambridge University Press.

Humber, W. (1857). *A practical treatise on cast and wrought iron bridges.* London.

Jeaffreson, J. S. (1864). *The life of Robert Stephenson.* London: Longman.

Jenkin, H. C. F. (1869). On the practical applications of reciprocal figures to the calculation of strains on framework, *Transactions of the Royal Society of Edinburgh,* **25,** 441–7.

Jenkin, H. C. F. (1873). On braced arches and suspension bridges, *Transactions of the Royal Scottish Society of Arts,* **8,** 135 (presented 13 December 1869).

Jourawski, D. J. (1856). Sur la résistance d'un corps prismatique et d'une pièce composée en bois ou en tôle de fer à une force perpendiculaire à leur longueur, *Mémoires Annales des Ponts et Chaussées,* sem. 2, 328–51 (presented to the Russian Academy of Sciences, 1854).

Koechlin, M. (1889*a*). *Les applications de la statique graphique.* Paris.

Koechlin, M. (1889*b*). La tour de 300 metres à l'exposition universelle de Paris, *Schweizerische Bauzeitung,* **13,** 23–8; **14,** 16–20.

Könen, F. (1882). Vereinfachung der Berechnung kontinuierlicher Balken mit Hilfe des Satzes von der Arbeit, *Wochenblatt für Architekten und Ingenieur,* **21,** 402.

Köpeke, E. (1856). Ueber die Dimensionen von Balkenlagen, *Zeitschrift des Architekten- und Ingenieur-Vereins zu Hannover,* **2,** cols. 82–91.

Kopytowski (1865), *Ueber die inneren Spannungen in einem Freiaufliegenden Balken unter einwirkung beweglicher Belastung.* Göttingen.

Krohn, R. (1884). Der Satz von der Gegenseitigkeit der Verschiebungen, und Anwendung desselben zur Berechnung statisch unbestimmte Fachwerkträger, *Zeitschrift des Architekten- und Ingenieur-Vereins zu Hannover,* **30,** col. 269–74.

La Hire, P. de (1695). *Traité de mécanique.* Paris.

Lamarle, E. (1855). Note sur un moyen très-simple d'augmenter, dans une proportion notable, la résistance d'une pièce prismatique chargée uniformément, *Bulletin de l'Académie Royale Belgique,* **22,** part 1, 232–52, 503–25.

Lamé, G. & Clapeyron, B. P. E. (1823). Mémoire sur la stabilité des vôutes, *Annales des Mines,* **8,** 789.

Lamé, G. & Clapeyron, B. P. E. (1826a). Mémoire sur l'emploi du fer dans les ponts suspendus, *Journal des Voies de Communication*, 5, 19–43; 9, 29–55.

Lamé, G. & Clapeyron, B. P. E. (1826b). Mémoire sur la construction des polygons funiculaires, *Journal des Voies de Communication*, 5, 63ff.

Lamé, G. & Clapeyron, B. P. E. (1831). Mémoire sur l'équilibre intérieur des corps solides homogènes, *Crelles Journal für die reine und angewandte Mathematik*, 7, 150–69, 237–52, 381–413 (submitted to the French Academy of Sciences (1828) and published (1832) in *Mémoires présentés par divers Savans*, 4, 465–562).

Lamé, G. (1852). *Leçons sur la théorie mathématique de l'élasticité des corps solides*. Paris: Gauthier–Villars.

Land, R. (1888). Kinematische Theorie der statisch bestimmten Träger, *Zeitschrifte des österreichischen Ingenieur und Architekten Vereins*, 40, 11–162.

Landsberg, H. (1885). Beitrag zur Theorie der Fachwerke, *Zeitschrift des Architekten- und Ingenieur-Vereins zu Hannover*, 31, col. 362; 32, col. 195.

Levy, M. (1874). *La statique graphique et ses applications aux constructions*. Paris: Gauthier–Villars. (part 2, 1886a.)

Levy, M. (1886). Mémoire sur le calcul des ponts suspendus rigides, *Annales des Ponts et Chaussées*, ser. 2, 12, 179–246.

Love, E. H. (1892). *A treatise on the mathematical theory of elasticity* (4th edn, 1927). Cambridge University Press.

Mach, E. (1883). *Die Mechanik in ihrer Entwickelung*. Leipzig: Brockhaus. (2nd edn, 1888.)

Mallet, R. (1860). On the increased deflection of girders on bridges exposed to the transverse strain of a rapidly passing load, *Civil Engineer and Architect's Journal*, 23, 109–10.

Manderla, H. (1880). Die Berechnung der Sekundärspannungen, *Allgemeine Bauzeitung*, 45, 34.

Martin, H. M. (1879). Arched ribs and voussoir arches. *Minutes of the Proceedings of the Institution of Civil Engineers*, 93, 462ff.

Martin, H. M. (1895). *Statically-indeterminate structures and the principle of least work*. London: Engineering.

Maxwell, J. C. (1864a). On reciprocal figures and diagrams of forces, *London, Edinburgh and Dublin Philosophical Magazine*, ser. 4, 27, 250.

Maxwell, J. C. (1864b). On the calculation of the equilibrium and stiffness of frames, *London, Edinburgh and Dublin Philosophical Magazine*, ser. 4, 27, 294.

Maxwell, J. C. (1867). On the application of the theory of reciprocal polar figures to the construction of diagrams of forces, *The Engineer*, 24, 402.

Maxwell, J. C. (1870a). On reciprocal figures, frames and diagrams of forces, *Proceedings of the Royal Society of Edinburgh*, 7, 53–6.

Maxwell, J. C. (1870b). On reciprocal figures, frames and diagrams of forces, *Transactions of the Royal Society of Edinburgh*, 26, 1–46.

Maxwell, J. C. (1877). History of the doctrine of energy. In *Matter and Motion*. London.

Melan, J. (1883). Berechnung eiserner Hallengespärre unterAnwendung des Satzes von der kleinsten Arbeit, *Wochenschrift des österreichischen Architekten und Ingenieur Vereins*, 35, 149, 162.

Melan, J. (1888). *Theorie der Eisenen Bogenbrücken und der Hangenbrücken* (2nd edn). Berlin.

Menabrea, L. F. (1858). Nouveau principe sur la distribution des tensions dans les systèmes élastiques, *Comtes Rendus hebdomadaires des Séances de l'Académie des Sciences*, Paris, 46, 1056–61.

Menabrea, L. F. (1884). Sur la concordance de quelques méthodes générales pour déterminer les tensions dans un système de points réunis par des liens élastiques et sollicités par des forces extérieures en équilibre, *Comptes Rendus hebdomadaires des Séances de l'Académie des Sciences*, Paris, **98**, 714–17.

Möbius, A. F. (1837). *Lehrbuch der Statik*. Berlin.

Mohr, O. C. (1860). Beiträge zur Theorie der Holz- und Eisenkonstruktionen, *Zeitschrift des Architekten- und Ingenieur-Vereins zu Hannover*, **6**, cols. 323, 407; 1862, **8**, col. 245; 1868, **14**, col. 19.

Mohr, O. C. (1870). Beitrag zur Theorie der elastischen Bogenträger, *Zeitschrift des Architekten- und Ingenieur-Vereins zu Hannover*, **16**, col. 389.

Mohr, O. C. (1874*a*). Beitrag zur Theorie der Bogenfachwerksträger, *Zeitschrift des Architekten- und Ingenieur-Vereins zu Hannover*, **20**, col. 223.

Mohr, O. C. (1874*b*). Beitrag zur Theorie des Fachwerks, *Zeitschrift des Architekten- und Ingenieur-Vereins zu Hannover*, **20**, col. 509; 1875, **21**, col. 17.

Mohr, O. C. (1881). Beitrag zur Theorie des Bogenfachwerks, *Zeitschrift des Architekten- und Ingenieur-Vereins zu Hannover*, **27**, col. 243.

Mohr, O. C. (1882). Über die darstellung des Spannungszustandes und des Deformationszustandes seines Körperelementes und über die Anwendung derselben in der Festigkeitslehre, *Der Civilingenieur*, **28**, col. 113–56.

Mohr, O. C. (1883). Ueber das sogenaunte Prinzip der kleinsten Deformationsarbeit, *Wochenblatt für Architekten und Ingenieur*, **24**, 171.

Mohr, O. C. (1885). Beitrag zur Theorie des Fachwerks, *Der Civilingenieur*, **31**, 289.

Mohr, O. C. (1887). Ueber Geschwindigkeitspläne und Beschleunigungspläne kinematischer Ketten, *Der Civilingenieur*, **33**, 631.

Mohr, O. C. (1892). Die Besechnung der Fachwerk mit starren Knotenverbindungen, *Der Civilingenieur*, **38**, 577; 1893, **39**, 67.

Mohr, O. C. (1906). *Abhandlungen aus dem Gebiete der technischen Mechanik*. Berlin: Wilhelm Ernst. (2nd edn, 1914.)

Molinos, L. & Pronnier, C. (1857). *Traité théoretique et pratique de la construction des ponts métalliques*. Paris.

Morin, L. (1853). *Leçons de mécanique pratique*: vol. 5, *Résistance des matériaux*. Paris.

Moseley, H. (1833*a*). On a new principle in statics, called the principle of least pressure, *London, Edinburgh and Dublin Philosophical Magazine*, ser. 3, **3**, 285–8.

Moseley, H. (1833*b*). On the theory of resistances in statics, *London, Edinburgh and Dublin Philosophical Magazine*, ser. 3, **3**, 431–6.

Moseley, H. (1835). On the equilibrium of the arch, *Cambridge Philosophical Transactions*, **5**, 293–313 (read, 9 December 1833).

Moseley, H. (1838). On the theory of the equilibrium of bodies in contact, *Cambridge Philosophical Transactions*, **6**, 463–91 (read, 15 May 1837).

Moseley, H. (1843). *The mechanical principles of engineering and architecture*. London: Longman.

Müller-Breslau, H. F. B. (1880). *Theorie und Berechnung der eisenen Bogenbrücken*. Berlin.

Müller-Breslau, H. F. B. (1881). Theorie der Durch einen Balken versteiften Kette, *Zeitschrift des Architekten- und Ingenieur-Vereins zu Hannover*, **27**, col. 61; 1883, **29**, col. 347.

Müller-Breslau, H. F. B. (1883). Zur Theorie der versteifung labiler flexibler Bogenträger, *Zeitschrift für Bauwesen*, **33**, 312.

Müller-Breslau, H. F. B. (1884*a*). Der Satz von der Abgeleiteten der idealen Formanderungsarbeit, *Zeitschrift des Architekten- und Ingenieur-Vereins zu Hannover*, **30**, col. 211.

Müller-Breslau, H. F. B. (1884*b*). Einflusslinien fur kontinuerliche Träger mit drei Stutzpunkten, *Zeitschrift des Architekten- und Ingenieur-Vereins zu Hannover*, **30**, col. 278 (also in *Wochenblatt für Architekten und Ingenieur*, **22** (1883), 353).

Müller-Breslau, H. F. B. (1884*c*), Vereinfachung der Theorie der statisch unbestimmten

Bogenträger, *Zeitschrift des Architekten- und Ingenieur-Vereins zu Hannover*, **30**, col. 575.

Müller-Breslau, H. F. B. (1885). Beitrag zur Theorie des Fachwerks, *Zeitschrift des Architekten- und Ingenieur-Vereins zu Hannover*, **31**, col. 417.

Müller-Breslau, H. F. B. (1886a). Zur Theorie der Biegungsspannungen in Fachwerkträgern, *Zeitschrift des Architekten- und Ingenieur-Vereins zu Hannover*, **32**, col. 399.

Müller-Breslau, H. F. B. (1886b). *Die neueren Methoden der Festigkeitslehre und der Statik der Baukonstruktionen*. Leipzig: Baumgartner.

Müller-Breslau, H. F. B. (1887a). Beitrag zur Theorie des ebenen Fachwerks, *Schweizerische Bauzeitung*, **9**, 121, 129.

Müller-Breslau, H. F. B. (1887b). *Die graphische Statik der Baukonstruktionen*, vol. 1; vol. 2, 1892. Leipzig: Kroner.

Müller-Breslau, H. F. B. (1888). Beitrag zur Theorie der ebenen elastischen Träger, *Zeitschrift des Architekten- und Ingenieur-Vereins zu Hannover*, **34**, col. 605.

Müller-Breslau, H. F. B. (1891). Über einige Aufgaben der Statik welche auf Gleichungen der Clapeyronschen art Führen, *Zeitschrift für Bauwesen*, **41**, 103–28.

Navier, C. L. M. H. (1823). *Rapport et mémoire sur les ponts suspendus* (2nd edn, 1830, including details of le Pont des Invalides and the effect of longitudinal impact on bars). Paris.

Navier, C. L. M. H. (1825). Note sur les questions de statique dans lesquelles on considère un corps pesant supporté par un nombre de points d'appui surpassant trois, *Nouveau Bulletin des Sciences par la Société Philomatique de Paris*, **11**, 35–7.

Navier, C. L. M. H. (1826). *Résumé des leçons données à l'Ecole des Ponts et Chaussées sur l'application de la mécanique à l'établissement des constructions et des machines*, 2nd edn, 1833. Paris: Carilian–Goeury.

Navier, C. L. M. H. (1864). *Résumé des leçons données à l'Ecole des Ponts et Chaussées sur l'application de la mécanique à l'établissement des constructions et des machines*, 3rd edn, with notes and appendices by B. de Saint-Venant.

Navier, C. L. M. H., see Gauthey (1809).

Navier, C. L. M. H., see Bélidor (1729).

Navier, C. L. M. H., see Bélidor (1737).

Niles, A. S. (1950). Clerk Maxwell and the theory of indeterminate structures, *Engineering*, **170**, 194–8.

Ovazza, E. (1888). Sul calcolo delle deformazione dei sistemi articolati, *Atti della Reale Accademia delle Scienze di Torino*, **23**, 62–9.

Owen, J. B. B. (1976). Arch Bridges. In *The works of Isambard Kingdom Brunel*, Alfred Pugsley (ed.). London: The Institution of Civil Engineers and The University of Bristol. (2nd edn, 1980. Cambridge University Press.)

Phillips, E. (1855). Calcul de la résistance des poutres droites telles que les ponts etc., sous l'action d'une charge en mouvement, *Annales des Mines*, **7**, 467–506 (summarised in *Comptes Rendus hebdomadaires des Séances de l'Académie des Sciences*, **46**, 30–2).

Poirée, J. (1854). Observations sur la séparation de la pression dans la section transversale des arcs des ponts en fonte, *Annales des Ponts et Chaussées*, sem. 1, 374–95.

Pippard, A. J. S. & Baker, J. F. (1938). *Analysis of engineering structures*, 3rd edn, 1957. London.

Pole, W. (1850). Strength and deflection of beams: continuous beams. In *Britannia and Conway tubular bridges*, E. Clark (ed.). London: Longman.

Pole, W. (1864). Iron bridges. In *The life of Robert Stephenson*, J. S. Jeaffreson. London: Longman.

Pole, W. (1877). *The Life of Sir William Fairbairn*. London: Longman.

Poncelet, J. V. (1822). *Traité des propriétés projectives des figures*. Paris.

Poncelet, J. V. (1826). *Cours de mécanique appliqué aux machines.* Paris.

Poncelet, J. V. (1831). *Cours de mécanique industrielle fait aux artistes et ouvriers messins,* part 3. Metz.

Poncelet, J. V. (1835). Mémoire sur la stabilité des revêtements et leur fondations, *Mémoires de l'Officier de Génie,* **12,** 36ff; **13,** 25ff.

Poncelet, J. V. (1852). Examen critique et historique des principales théories ou solutions concernant l'équilibre des voûtes, *Comptes Rendus hebdomadaires des Séances de l'Académie des Sciences,* **35,** 494, 531, 577.

Pugsley, A. G. (1957). *The theory of suspension bridges,* 2nd edn, 1968. London: Arnold.

Pugsley, A. G. (1976). (ed.) *The works of Isambard Kingdom Brunel.* London: The Institution of Civil Engineers and The University of Bristol. (2nd edn, 1980. Cambridge University Press.)

Rankine, W. J. M. (1858). *A manual of applied mechanics.* London: Griffin.

Rankine, W. J. M. (1861). *A manual of civil engineering.* London: Griffin.

Rankine, W. J. M. (1864). Principle of the equilibrium of polyhedral frames, *London, Edinburgh and Dublin Philosophical Magazine,* ser. 4, **27,** 92.

Rankine, W. J. M. (1870). Diagrams of forces in frameworks, *Proceedings of the Royal Society of Edinburgh,* **7,** 171–2.

Rankine, W. J. M. (1872). Frames of roofs, *The Engineer,* **33,** 114.

Rayleigh (J. W. Strutt) Baron. (1873). Some general theorems relating to vibrations, *Proceedings of the London Mathematical Society,* **4,** 357–68.

Rayleigh (J. W. Strutt) Baron. (1874). A statical theorem, *London, Edinburgh and Dublin Philosophical Magazine,* ser. 4, **48,** 452; **49,** 183.

Rayleigh (J. W. Strutt) Baron. (1877). *Theory of sound,* 2 vols. London: Macmillan.

Rebhann, G. (1856). *Theorie der Holz und Eisen-constructionen.* Vienna.

Renaudot, A. (1861). Etude de l'influence des charges en mouvement sur la résistance des ponts métalliques à poutres droites, *Annales des Ponts et Chaussées,* ser. 4, **1,** 145–204.

Reuleaux, F. (1865). *Der Constructeur.* Brunswick.

Reye, W. (1868). *Geometrie der Lage.* Hannover.

Ritter, A. (1862). *Elementare Theorie der Dach und Brückenconstructionen,* 3rd edn, 1879. Hannover.

Ritter, W. (1877). Theorie und Berechnung der Eisenbrucken, *Zeitschrift für Bauwesan,* **27,** 189.

Ritter, W. (1893). *Die elastische Linie und ihre Anwendung auf den continuirlichen Balken.* Zurich: Meyer & Zeller.

Ritter, W. (1884). Über die Druckfestigkeit stabförmiger Körper mit besonderer Rücksicht auf die ein steiften Fachwerk auftretenden Nebenspannungen, *Schweizerische Bauzeitung,* **1,** 37, 43, 47.

Ritter, W. (1886). *Der elastische Bogen berechnet mit Hilfe der graphische Statik.* Zurich: Raustein.

Ritter, W. (1888–1907). *Anwendungen der graphischen Statik,* 4 vols. Zurich: Raustein.

Rohrs, J. H. (1856). On the oscillations of a suspension chain, *Cambridge Philosophical Transactions,* **9,** 379–98 (read, December 1851).

Rouse, H. & Ince, S. (1957). *A history of hydraulics.* U.S.A.: Iowa.

Saint-Venant, B. de (1857). Sur l'impulsion transversale et la résistance vive des barres élastiques appuyées aux extrémités, *Comptes Rendus hebdomadaires des Séances de l'Académie des Sciences,* Paris, **45,** 204–8.

Saint-Venant, B. de (1867). Sur le choc longitudinal de deux barres élastiques, *Journal de Liouville,* **12,** 237–376.

Saint-Venant, B. de, *see* Navier (1864).

Saint-Venant, B. de, *see* Clebsch (1862).

Salmon, E. H. (1938). *Materials and structures,* 2 vols. London: Longman.

Saviotti, C. (1875). La statica grafica, *Atti dell'Accademia Nazionale dei Lincei*, Rome, part 3, **2**, 148.

Saviotti, C. (1888). *La statica grafica*. Milan.

Scheffler, H. (1857). *Theorie Gewolbe Futtermauern und eisenen Brucken*. Brunswick.

Scheffler, H. (1858a). Festigkeits- und Biegungsverhältnisse eines über mehrere Stützpunkte fortlautenden Trägers, *Der Civilingenieur*, **4**, 62–73.

Scheffler, H. (1858b). Continuirliche Brückenträger, *Der Civilingenieur*, **4**, 142–6; 1860, **6**, 129–202.

Scheffler, H. (1858c). *Theorie der Festigkeit gegen das Zerknicken nebst Untersuchungen über verschiedenen inneren Spannungen gebogener Körper und über andere Probleme der Biegungstheorie mit praktischen Anwendungen*. Brunswick.

Schur, F. (1895). Über ebene einfache Fachwerke, *Zeitschrift für Mathematik und Physik*, **40**, 48ff.

Schwedler, J. W. (1851). Theorie der Bruckenbalkensysteme, *Zeitschrift für Bauwesen*, **1**, 114, 162, 265.

Skilbinski, K. (1883). Das Deformations Polygon und dessen Anwendungen zur graphischen Berechnung statisch unbestimmte Fachwerke, *Zeitschrift des österreichischen Ingenieur und Architekten Vereins*, **40**, 23.

Snell, G. (1846). On the stability of arches and practical methods for determining, according to the pressures to which they will be subjected, the best form of section or variable depth of voussoir for any given intrados or extrados, *Minutes of the Proceedings of the Institution of Civil Engineers*, **5**, 439–74.

Stokes, G. G. (1849). Discussion of a differential equation relating to the breaking of railway bridges, *Cambridge Philosophical Transactions*, **8**, 707.

Straub, H. (1952). *A history of civil engineering*, London: Leonard Hill. (First published (1949). *Die Geschichte der bauingenieurkunst*. Basel.)

Stussi, F. (1951). Centenary Rectorial Address, E.T.H., Zurich, *Schweizerische Bauzeitung*, **69**, 21–7.

Swain, G. F. (1882). Mohr's graphical theory of earth pressure, *Journal of the Franklin Institute*, **114**, 241–51.

Swain, G. F. (1883). On the application of the principle of virtual velocities to the determination of the deflections and stresses of frames, *Journal of the Franklin Institute*, **115**, 102.

Tait, P. G. (1864). A note on the history of energy, *London, Edinburgh and Dublin Philosophical Magazine*, ser. 4, **28**, 55.

Tait, P. G. (1868). *Sketch of thermodynamics: historical sketch of the science of energy*. Edinburgh: Edmonston and Douglas.

Tellkampf, H. (1856). *Die Theorie der Hängbrücken mit Besonderer Rücksicht auf deren anwendung*. Hannover.

Tetmajer, L. (1882). *Culmann's enduring achievements*. Zurich.

Thomson, W. (1857). On the thermo-elastic and thermo-magnetic properties of matter, *Quarterly Journal of Mathematics*, **1**, 57–77.

Thomson, W. & Tait, P. G. (1879). *Treatise on natural philosophy*. Cambridge University Press.

Timoshenko, S. P. (1950). *D. J. Jourawski and his contribution to the theory of structures*. Vienna: Federhöfer–Girkmann.

Timoshenko, S. P. (1953). *History of strength of materials*. New York: McGraw-Hill.

Todhunter, I. & Pearson, K. (1886–1893). *A history of the theory of elasticity and of the strength of materials from Galilei to the present time*, 3 vols. Cambridge University Press.

Unwin, W. (1869). *Wrought iron bridges and roofs*. London.

Varignon, P. (1725). *Nouvelle mécanique ou statique dont le projet fut donné en 1687* (written, 1687). Paris.

Vierendeel, M. (1912). *Cours de stabilité des constructions.* Louvain.

Villarceau, L. Yvon. (1853): *Sur l'établissement des arches de pont.* Paris.

Weale, J. (1843). (ed.). *The theory, practice and architecture of bridges of stone, iron, timber and wire with examples on the principle of suspension.* London.

Westergaard, H. M. (1930). One hundred and fifty years advance in structural analysis, *Transactions of the American Society of Civil Engineering,* **94,** 226–40 (presented 8 October 1926).

Weyrauch, J. J. (1873). *Allgemeine Theorie der kontinuerlichen und einfachen Träger.* Leipzig: Teubner.

Weyrauch, J. J. (1874). *Über die graphische Statik.* Leipzig: Teubner.

Weyrauch, J. J. (1878). Theorie der elastischen Bogenträger, *Zeitschrift für Baukunde,* **31,** 367.

Weyrauch, J. J. (1880). On the calculation of dimensions as depending on the ultimate working strength of materials. *Minutes of the Proceedings of the Institution of Civil Engineers,* **63,** 275–96; **71** (1881), 298.

Weyrauch, J. J. (1884). *Theorie elastische Körper: eine Einleitung zur Mathematischen Physik und technischen Mechanik.* Leipzig: Teubner.

Weyrauch, J. J. (1885). *Aufgaben zur Theorie elastische Körper.* Leipzig: Teubner.

Weyrauch, J. J. (1886). Arbeitsbedingungen für statisch unbestimmte Systeme, *Wochenblatt für Architekten und Ingenieur,* **38,** 200.

Weyrauch, J. J. (1887). *Theorie der statisch bestimmten Träger.* Leipzig: Teubner.

Weyrauch, J. J. (1888). *Beispiele und Aufgaben zur Berechnung der statisch bestimmten Träger.* Leipzig: Teubner.

Weyrauch, J. J. (1896). *Elastische Bogenträger.* Munich.

Whewell, W. (1834). *Mechanics applied to the arts.* Cambridge University Press.

Whewell, W. (1841). *The mechanics of engineering* (dedicated to R. Willis). Cambridge University Press.

Whipple, S. (1847). An essay on bridge building. *Elementary and practical treatise on bridge building,* 2nd edn, 1873. New York.

Williamson, B. (1894). *Stress and strain of elastic solids.* London: Longman.

Williot, M. (1877a). Notions pratiques sur statique graphique, *Annales des Génies Civils,* **6,** 601.

Williot, M. (1877b). *Notions pratiques sur la statique graphique.* Paris.

Willis, R., James, H. & Galton, D. (1849). Experiments for determining the effects produced by causing weights to travel over bars with different velocities. *Report of the Commissioners appointed to inquire into the application of iron to railway structures* (appendix B). London.

Wilson, G. (1897). On a method of determining the reactions at the points of support of a continuous beam, *Proceedings of the Royal Society of London,* **62,** 268–77.

Winkler, E. (1858). Formänderung und Festigkeit gekrümmer Körper insbesondere derringe, *Der Civilingenieur,* **40,** 232–46.

Winkler, E. (1860). Die inneren Spannungen deformirter insbesondere auf relative Festigkeit in Anspruch genomener Körper, *Zeitschrift für Bauwesen,* **10,** 93–108, 22–36, 265–80.

Winkler, E. (1862). Beiträge zur Theorie der kontinuerlichen Brückenträger. *Der Civilingenieur,* **8,** 135–82.

Winkler, E. (1867). *Die Lehre von der Elastizität und Festigkeit.* Prague.

Winkler, E. (1868). Vortrag über die Berechnung von Bogenbrucken, *Mittheilungen des Architekten- und Ingenieur-Vereins in Böhmen,* **7,** 6 (on influence lines).

Winkler, E. (1872). *Vorträge ubër Brückenbau, Theorie der Brücken:* vol. 1, 2nd edn, 1875, *Aeussere Kräfte gerader Träger.* Carl Gerald's Sohn.

Winkler, E. (1873). *Neue Theorie des Erddruckes.* Vienna.

Winkler, E. (1879a). Beitrag zur Theorie der elastischen Bogenträger. *Zeitschrift des Architekten- und Ingenieur-Vereins zu Hannover,* **25**, 199.

Winkler, E. (1879b). Lage der Stutzlinie im Gewolbe, *Deutsche Bauzeitung,* **44**, 117; **45** (1880a), 58.

Winkler, E. (1880). Beitrag zur Werthschätzung der Träger mit statisch bestimmten mehrtheiligem Gitterwerke, *Wochenblatt für Architekten und Ingenieur,* **32**, 162ff.

Winkler, E. (1881a). *Vorträge uber Bruckenbau, Theorie der Brucken:* vol. 2, *Innere Krafte gerader Träger.* Vienna: Carl Gerald's Sohn.

Winkler, E. (1881b). Die sekundärspannungen in eisenkonstruktionen, *Deutsche Bauzeitung,* **46**, 72–9.

Young, A. E. (1898). Rankine's treatment of the elastic arch, *Minutes of the Proceedings of the Institution of Civil Engineers,* **131**, 323–37.

Young, T. (1807). *A course of lectures on natural philosophy and the mechanical arts.* London.

Name index

Subject index

192   Subject index

Printed in the United States
By Bookmasters